T0210541

SpringerBriefs in Electrical and Computer
Engineering

More information about this series at http://www.springer.com/series/10059

Haibo Zhou • Lin Gui • Quan Yu
Xuemin (Sherman) Shen

Cooperative Vehicular Communications in the Drive-thru Internet

Springer

Haibo Zhou
Department of Electrical
   and Computer Engineering
University of Waterloo
Waterloo, ON, Canada

Quan Yu
Department of Electrical Engineering
Shanghai Jiao Tong University
Shanghai, China

Lin Gui
Department of Electrical Engineering
Shanghai Jiao Tong University
Shanghai, China

Xuemin (Sherman) Shen
Department of Electrical
   and Computer Engineering
University of Waterloo
Waterloo, ON, Canada

ISSN 2191-8112   ISSN 2191-8120 (electronic)
SpringerBriefs in Electrical and Computer Engineering
ISBN 978-3-319-20453-6   ISBN 978-3-319-20454-3 (eBook)
DOI 10.1007/978-3-319-20454-3

Library of Congress Control Number: 2015943619

Springer Cham Heidelberg New York Dordrecht London

Printed on acid-free paper

Springer International Publishing AG Switzerland is part of Springer Science+Business Media (www.springer.com)

Preface

Driven by the ever-growing expectation of ubiquitous connectivity and widespread adoption of IEEE 802.11 networks, it is highly demanded for in-motion vehicles to establish convenient Internet access to the roadside Wi-Fi access points (APs), which is referred to as Drive-Thru Internet. Due to the high vehicle mobility, harsh and intermittent wireless channels, limited APs coverage, and instinct issues of IEEE 802.11 MAC as it was originally designed for static scenarios, the wireless access performance and service quality of upper-layer applications, such as file download and video streaming, etc., in Drive-thru Internet would be severely restricted. On addressing those issues, cooperative vehicular communications have been leveraged as effective approaches for the performance enhancement in Drive-thru Internet and have attracted an extensive research attentions currently. In this monograph, we first provide an overview of the current state-of-the-art research works on the Drive-thru Internet. Then, to investigate the Drive-thru Internet performance in the outdoor vehicular environment considering the effect of throughput performance anomaly and a large number of fast-motion contending vehicles, a unified Drive-thru Internet analytical framework is proposed, which can accommodate various vehicular traffic flow states and be compatible with IEEE 802.11 networks with distributed coordination function (DCF). Accordingly, a spatial access control management approach is introduced, which has been verified a boosted throughput performance of Drive-thru Internet in a practical, efficient, and distributed manner. Further, we present ChainCluster, a cooperative Drive-thru Internet scheme by adopting a delicate linear cluster formation approach to support the flexible and efficient vehicular content distribution in Drive-thru Internet. Using simulations, the performance of ChainCluster has been shown to outperform that of the existing clustering schemes. Finally, we end the monograph by outlining some future research directions for cooperative vehicular communications in Drive-thru Internet.

We would like to thank Prof. Fen Hou, Prof. Tom H. Luan, Ning Zhang, Ning Lu, Miao Wang, Nan Cheng, and Khadige Abboud from Broadband Communications Research Group (BBCR) at the University of Waterloo, Prof. Bo Liu and Jiacheng Chen from Shanghai Jiaotong University, and Dr. Fan Bai from General Motor, for their contributions in the presented research works. Special thanks are also due to the staff at Springer Science+Business Media: Jennifer Malat and Melissa Fearon, for their help throughout the publication preparation process.

Waterloo, ON, Canada   Haibo Zhou
Shanghai, China   Lin Gui
Shanghai, China   Quan Yu
Waterloo, ON, Canada   Xuemin (Sherman) Shen

Contents

Acronyms

VANET	Vehicular ad-hoc networks
WLAN	Wireless local area networks
RSU	Road side unit
OBU	On-board unit
HTTP	Hypertext Transfer Protocol
IP	Internet Protocol
SMTP	Simple Mail Transfer Protocol
POP3	Post Office Protocol-version 3
DCF	Distributed coordination function
MAC	Media access control
V2I	Vehicle-to-infrastructure
V2R	Vehicle-to-roadside
V2V	Vehicle-to-vehicle
ITS	Intelligent Transportation System
DSRC	Dedicated short range communications
GPS	Global Positioning System
QoS	Quality of service
BS	Base Station
ARQ	Automatic Repeat Quest
CSMA	Carrier sense multiple access
CA	Collision avoidance
UWB	Ultra-Wide Band
SNR	Signal-to-noise ratio
FTM	Fluid traffic motion
CDF	Cumulative distributed function
PDF	Probability distributed function
RTS	Request to Send
CTS	Clear to Send
TCP	Transmission Control Protocol
UDP	User Datagram Protocol
ISM	Industrial Scientific Medical

Chapter 1
Introduction

The advance of wireless communications and pervasive use of mobile electronics in the recent years have driven the ever-increasing user demands, especially the bandwidth-hungry wireless demands, for the ubiquitous Internet access. This is particularly evident for in-vehicle Internet access with people now spending much their time in cars. By deploying a multitude of wireless gateway points along the roadside, namely roadside units (RSUs), the vehicles are possible to acquire temporary and opportunistic wireless connections to the Internet when driving through the coverage of RSUs, and accordingly enjoy the limited Internet services. Such a fundamental vehicular communication framework is referred to as the Drive-thru Internet. In this monograph, the RSU is also referred to as access point (AP). Therefore, we use RSU and AP interchangeably throughout the monograph. Compared with the Internet access services via ubiquitous connected cellular networks, Drive-thru Internet can form cost-effective, high-quality, and flexible vehicular communication networking by utilizing the technically matured Wi-Fi technology and widely deployed wireless local area networks (WLANs) communication infrastructure. The remainder of this chapter is organized as follows. Section 1.1 introduces the research background. Section 1.2 describes the basic concepts. Sections 1.3 and 1.4 discuss the challenges and contributions, respectively. Section 1.5 introduces the outline of this monograph.

1.1 Monograph Background

Vehicular ad hoc networks (VANET) have been widely recognized as an important enabling technology to implement a myriad of vehicular tailor-made applications ranging from the road safety, traffic management to vehicular infotainment [1]. Efficient and reliable Internet access services for VANET is a necessity for many

© The Author(s) 2015
H. Zhou et al., *Cooperative Vehicular Communications in the Drive-thru Internet*,
SpringerBriefs in Electrical and Computer Engineering,
DOI 10.1007/978-3-319-20454-3_1

vehicular applications, such as traffic information/social content publishing [2]. Typically, there are two types of Internet access approaches for in-motion vehicles, i.e., cellular networks and WLAN technique. However, both the two types of widely investigated vehicular Internet access approaches have their own individual deficiency or await further improvement and perfection. For instance, cellular networks with ubiquitous coverage strive for enabling ubiquitous communications by building up extensive base station infrastructures which are run by mobile phone operators, and can support continuous network access with lower bit-rates by convenient mobile access environment. However, cellular technique is debatable for vehicular applications due to the scarce licensed spectrum and high charges of cellular service data. Meanwhile, the current IEEE 802.11 standard based WLAN technology with very low cost is doubtable for in-motion vehicular Internet access due to the intermittent connectivity and comparable low drive-thru throughput which is caused by the high vehicular mobility, severe channel contentions, and particularly the limited coverage range. With the predictable booming of Internet-integrated vehicle services, especially the media-rich vehicular Internet access services on the go, an extensive body of researches have been devoted to enabling the cost-effective, high-quality, and flexible vehicular communication networking.

1.2 Basic Concepts

The recent advances in wireless communication techniques have made it possible for fast-moving vehicles to download data from the WLAN-based infrastructure, i.e., IEEE 802.11 technique based RSU. By deploying a multitude of WLAN-based wireless gateway points along the roadside, the vehicles are possible to acquire temporary and opportunistic wireless connections to the Internet when driving through the coverage of RSUs, and accordingly enjoy limited Internet services, which is referred to as the Drive-thru Internet [3, 4]. As the given Drive-thru Internet highways scenario shown in Fig. 1.1, Drive-thru Internet systems are multiple-access wireless networks in which both the vehicular on-board units (OBU) and people with wireless equipments in moving vehicles can connect to the RSU to obtain Internet connectivity for some period of time as the vehicles pass through the RSU's coverage range.

To enable effective vehicular-to-RSU communications in the fast driving environments with "extreme" high mobility and intermittent connectivity, the authors Jorg Ott et al., propose a drive-thru architecture which comprehensively considers the connection interruptions, users' mobility, and IP stack reconfiguration, and introduces a connectivity management services above the transport layer [3, 5, 6]. The Drive-thru architecture is shown in Fig. 1.2, which is composed of the following three elements: drive-thru client, drive-thru proxy, and mobile application peers.

- Drive-thru client, which can act as an application layer gateway for HTTP, SMTP, and POP3 application layer protocols to the keep a persistent connection.

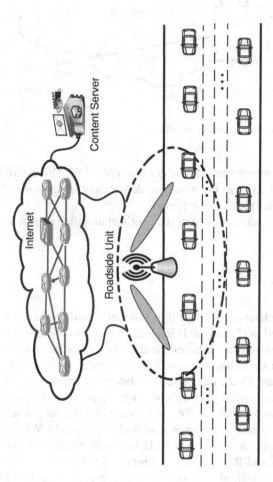

Fig. 1.1 The Drive-thru Internet highways scenario

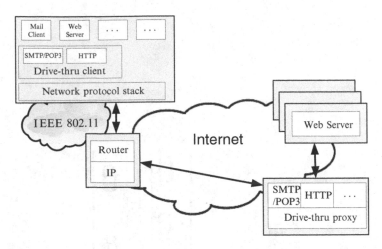

Fig. 1.2 The Drive-thru Internet architecture [5]

- Drive-thru proxy, which can play the counter-part of the Drive-thru client in the fixed network and conceals the mobile node's temporary unavailability from the corresponding (fixed) application peers.
- Mobile application peers, which are the user's unmodified standard applications.

1.3 Challenges

Despite that the effectiveness of Drive-thru Internet in real-world test environments has been well verified in [3] and the IEEE 802.11 distributed coordination function (DCF) has also been proven to be efficient to self-organize a mass of wireless nodes for fair dynamic medium access and resource sharing in both theory and real-world deployment [7–9], the Drive-thru Internet performance in the outdoor vehicular environment is still unclear when a large number of fast-motion nodes contend for the medium access at the same time [10]. One of the major concerns is the performance anomaly phenomenon in multi-rate 802.11 WLANs. As shown in Fig. 1.3, the transmission rate from RSU to a vehicle depends on the distance between the vehicle and RSU. When a number of vehicles contend the medium using the carrier sense multiple access with collision avoidance (CSMA/CA) and a vehicle located at the zone 4 obtains the current frame transmission opportunity, the transmission rate is 1 Mbps. With this low transmission rate, the time duration to transmit a frame by RSU will be longer than that for a vehicle located at the zone 1, which has a higher transmission rate. Therefore, the low transmission rate will greatly reduce the average throughput over all vehicles in the RSU's coverage. This phenomenon is referred as the performance anomaly.

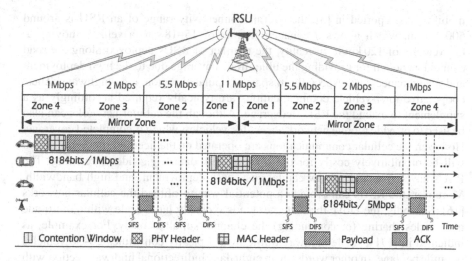

Fig. 1.3 The illustration of performance anomaly phenomenon in multi-rate 802.11b WLANs for vehicular medium access control

Due to the performance anomaly phenomenon in multi-rate 802.11 WLAN, the throughput of all nodes transmitting at the higher rate is degraded below the level of the lower rate, which penalizes high-rate nodes and privileges the slow-rate ones [7, 11]. To cope with this issue, the proposed solutions for static wireless users in WLANs both address the issue of fairness constraint and throughput optimization tradeoff [12, 13]. However, in the mobile vehicular environment, the throughput of Drive-thru Internet can be jointly determined by the following three perspectives: (1) vehicular related traffic flow state (e.g., vehicular density, driving speed); (2) IEEE 802.11 physical layer related fault parameters [14] (e.g., bit rate, coverage range); (3) IEEE 802.11 MAC layer related fault parameters (e.g., MAC frame size, contention window size). To address the performance anomaly in large-scale deployment of Drive-thru Internet, there are three challenges in efficiently adapting the IEEE 802.11 protocols according to the real-time road traffic with light computation cost and simple operations:

1. How to theoretically and accurately evaluate the IEEE 802.11 DCF throughput considering different traffic flow states and different versions of 802.11 products?
2. How to achieve the optimal IEEE 802.11 DCF throughput under different traffic flow states by overcoming the performance anomaly phenomenon in vehicular environment?
3. How to ensure the airtime fairness for medium sharing and boost the throughput performance of Drive-thru Internet?

Moreover, compared with the indoor WLAN scenarios, efficient communications in Drive-thru Internet is a much more challenging task [3, 4, 15], which mainly attributes to the following features of vehicular communications. Firstly, the wireless connectivity from vehicles to RSUs is transitory due to the high vehicle mobility. As reported in [3], the overall connectivity range of an RSU is around 500–600 m, which allows a connection time of 15–18 s to a vehicle moving at the velocity of 120 km/h. In reality, the number of RSUs deployed along the road cannot be enough for providing the ubiquitous coverage due to the high deployment and maintenance cost, especially in sparse populated region, e.g., highways. Thus, cooperative inter-vehicle communications are typically required accordingly as a supplement to extend the RSU's coverage. Although the cellular networks can also be used to provide mobile network access for vehicles, which is completed coverage. However, the cellular communications are operated on the licensed spectrum, which will be prohibitively costly for transferring bulk data, e.g., video streaming. The RSUs have advantage of low capital cost, easy deployment, and high bandwidth. Moreover, the RSUs can provide a potential way to offload the cellular networks [16]. Secondly, the Drive-thru Internet tends typically large-scale with a multitude of vehicles sharing (or contending) the channel simultaneously. For example, as indicated in [17] a stable highway traffic flow typically constitutes 20–30 vehicles per mile per lane. In other words, in an eight-lane bidirectional highway section with smooth traffic flow and vehicle-to-vehicle (V2V) communication range to be around 300 m, approximately 30–45 vehicles will share the channel for transmissions at the same time. Ott and Kutscher [3] reports a real-world measurement of the Drive-thru Internet on a highway road section. With a single vehicle connecting to a roadside IEEE 802.11b AP, a volume of 9 MB data per drive-thru can be acquired using either TCP or UDP at a vehicle velocity of 80 km/h. This allows the download of a medium-sized file, such as an MP3 file. However, when a multitude of vehicles share the RSU for the transmissions synchronously, the individual throughput performance would degrade significantly due to the transmission contentions and collisions and can hardly afford the upper-layer applications [10]. Therefore, a practical cooperative mechanism is highly demanded to convert vehicles from channel competitors to collaborators and enhance the welfare of all vehicles [18].

1.4   Highlights

This monograph focuses on the up-to-date researches on the cooperative vehicular communications in Drive-thru Internet, which covers the medium access control and vehicular content distribution related research issues. Followed by a comprehensive review and in-depth discussion of the current state-of-the-art research literature, this monograph presents a unified analytical framework for the Drive-thru Internet performance evaluation and accordingly proposes an optimal spatial access control management approach. The proposed spatial coordinated medium sharing approach can ensure the airtime fairness of medium sharing and boost the throughput performance of Drive-thru Internet in a practical, efficient, and distributed manner. To provide a flexible and efficient vehicular content distribution in Drive-thru Internet, this monograph also introduces a novel cooperative vehicular communication framework together with a delicate linear cluster formation scheme and low-delay content forwarding approach. We believe that our presented medium access control and vehicular content distribution related research results in this monograph will provide useful insights for the approaches design of WiFi technologies enabled vehicular communications and motivate a new line of thinking for the performance enhancements of future vehicular networking. Specifically, the highlights of the monograph are summarized as follows:

- Unified Drive-thru Internet analytical framework: In Chap. 3, existing literature largely adopts one of the IEEE 802.11 protocols (i.e., IEEE 802.11a/b/g) for the Drive-thru Internet performance analysis and evaluation. In this chapter, we derive the mathematical expression of the mean saturated throughput of Drive-thru Internet and accordingly the mean transmitted data volume per drive-thru, based on the widely deployed three types of network models (IEEE 802.11a/b/g networks) and fluid traffic motion model. Using empirical data, the proposed analytical framework is much closer to real vehicular environments and can be applied to analyze the Drive-thru Internet performance under various vehicular traffic flow states.
- Proposed optimal access control scheme: In Chap. 3, based on the collected empirical data from real-world WiFi networks and traffic transport environments, we develop an optimal access control scheme. The proposal optimally determines the best spatial region in which vehicles should start medium access and sharing to achieve the maximal system throughput. The proposed access control scheme is directly applicable to off-the-shelf IEEE 802.11a/b/g APs and has the advantage of simple and flexible operability. Moreover, through both analysis and extensive simulation results, we show that our proposal can ensure the airtime fairness for medium sharing and boost the throughput performance of Drive-thru Internet.
- Proposed linear cluster topology for vehicular cooperation: In Chap. 4, existing works largely group vehicles in a cluster where vehicles are mutually connected [19–21]. In contrast, we form cluster vehicles in a linear chain topology. The linear cluster is not only stable and allows reliable cooperations among cluster members, but also effective to extend the connection time of a single vehicle to that of multiple vehicles, and therefore boost the download performance.
- Well-studied fine-grain mobility model: In Chap. 4, existing literature largely adopts the macroscopic mobility [8, 22], which considers vehicles as traffic flows and typically evaluates the averaged performance. Such a method is not accurate enough to evaluate the download performance of a single content file in a specific drive-thru scenario with given neighboring vehicles and velocity. To address this issue, we apply the microscopic mobility model which captures the mobility features of a vehicle on highways to provide more accurate evaluation.
- Theoretical analysis and performance evaluation: In Chap. 4, we analyze the collective download performance of vehicles in the linear cooperation cluster on highways. In specific, we first derive the MAC throughput performance of vehicles with specific mobility features during the drive through of RSU. We then evaluate the integrated data volume that can be cooperatively downloaded by the

linear cluster. Finally, we analyze the downloads forwarding time to retrieve the whole content in the chain cluster. The proposed approach can provide systematic solution for the adaptive control of upper-layer media applications in highways VANET, such as video streaming.

1.5 Outline of the Monograph

The remainder of this monograph is organized with five chapters: In this chapter, we introduce the research background of cooperative Drive-thru Internet, and accordingly the basic concepts, research challenges, and monograph Highlights. In Chap. 2, we provide an overview of the current state-of-the-art research literature about the vehicular communications in Drive-thru Internet. In Chap. 3, we investigate the Drive-thru Internet performance in the outdoor vehicular environment considering the effect of performance anomaly phenomenon and different traffic states. Accordingly, a spatial access control management approach is introduced as well. In Chap. 4, we present ChainCluster, a cooperative Drive-thru Internet scheme by adopting a delicate linear cluster formation approach to support the flexible and efficient vehicular content distribution in Drive-thru Internet. In Chap. 5, we provide our view on further research issues for Drive-thru Internet.

References

1. H. Hartenstein, K.P. Laberteaux, A tutorial survey on vehicular ad hoc networks. IEEE Commun. Mag. **46**(6), 164–171 (2008)
2. T. Luan, X. Shen, F. Bai, L. Sun, Feel bored? Join verse! engineering vehicular proximity social network. IEEE Trans. Veh. Technol. **64**(3), 1120–1131 (2015)
3. J. Ott, D. Kutscher, Drive-thru Internet: IEEE 802.11 b for "Automobile" users, in *Proceedings of IEEE INFOCOM*, 2004, pp. 362–373 (2004)
4. N. Lu, N. Zhang, N. Cheng, X. Shen, J.W. Mark, F. Bai, Vehicles meet infrastructure: toward capacity-cost tradeoffs for vehicular access networks. IEEE Trans. Intell. Transp. Syst. **14**(3), 1266–1277 (2013)
5. J. Ott, D. Kutscher, The "drive-thru" architecture: Wlan-based internet access on the road, in *Proceedings of VTC*, vol. 5 (IEEE, Los Angeles, 2004), Los Angeles, CA, pp. 2615–2622
6. J. Ott, D. Kutscher, A disconnection-tolerant transport for drive-thru internet environments, in *Proceedings of INFOCOM*, vol. 3 (IEEE, Miami, 2005), Miami, pp. 1849–1862
7. M. Heusse, F. Rousseau, G. Berger-Sabbatel, A. Duda, Performance anomaly of 802.11 b, in *Proceedings of INFOCOM*, vol. 2 (IEEE, San Francisco, 2003), San Francisco, pp. 836–843
8. W.L. Tan, W.C. Lau, O. Yue, T.H. Hui, Analytical models and performance evaluation of drive-thru Internet systems. IEEE J. Sel. Areas Commun. **29**(1), 207–222 (2011)
9. L. Dai, X. Sun, A unified analysis of ieee 802.11 dcf networks: stability, throughput and delay. IEEE Trans. Mob. Comput. **12**(8), 1558–1572 (2013)
10. T.H. Luan, X. Ling, X. Shen, Mac in motion: impact of mobility on the mac of drive-thru internet. IEEE Trans. Mob. Comput. **11**(2), 305–319 (2012)

11. D.B. Rawat, D.C. Popescu, G. Yan, S. Olariu, Enhancing vanet performance by joint adaptation of transmission power and contention window size. IEEE Trans. Parallel Distrib. Syst. **22**(9), 1528–1535 (2011)
12. D.-Y. Yang, T.-J. Lee, K. Jang, J.-B. Chang, S. Choi, Performance enhancement of multirate ieee 802.11 wlans with geographically scattered stations. IEEE Trans. Mob. Comput. **5**(7), 906–919 (2006)
13. Y.-L. Kuo, K.-W. Lai, F.-S. Lin, Y.-F. Wen, E.-K. Wu, G.-H. Chen, Multirate throughput optimization with fairness constraints in wireless local area networks. IEEE Trans. Veh. Technol. **58**(5), 2417–2425 (2009)
14. P. Belanovic, D. Valerio, A. Paier, T. Zemen, F. Ricciato, C.F. Mecklenbrauker, On wireless links for vehicle-to-infrastructure communications. IEEE Trans. Veh. Technol. **59**(1), 269–282 (2010)
15. T.H. Luan, X. Ling, X. Shen, Provisioning QoS controlled media access in vehicular to infrastructure communications. Ad Hoc Netw. **10**(2), 231–242 (2012)
16. K. Lee, J. Lee, Y. Yi, I. Rhee, S. Chong, Mobile data offloading: how much can wifi deliver? IEEE/ACM Trans. Netw. **21**(2), 536–550 (2013)
17. A.D. May, *Traffic Flow Fundamentals* (Prentice Hall, Englewood Cliffs, 1990)
18. W. Zhuang, M. Ismail, Cooperation in wireless communication networks. IEEE Wirel. Commun. **19**(2), 10–20 (2012)
19. H. Su, X. Zhang, Clustering-based multichannel mac protocols for qos provisionings over vehicular ad hoc networks. IEEE Trans. Veh. Technol. **56**(6), 3309–3323 (2007)
20. Z. Wang, L. Liu, M. Zhou, N. Ansari, A position-based clustering technique for ad hoc intervehicle communication. IEEE Trans. Syst. Man Cybern. C Appl. Rev. **38**(2), 201–208 (2008)
21. Y.-C. Lai, P. Lin, W. Liao, C.-M. Chen, A region-based clustering mechanism for channel access in vehicular ad hoc networks. IEEE J. Sel. Areas Commun. **29**(1), 83–93 (2011)
22. Y. Zhuang, J. Pan, V. Viswanathan, L. Cai, On the uplink Mac performance of a drive-thru internet. IEEE Trans. Veh. Technol. **61**(4), 1925–1935 (2012)

Chapter 2
Overview of Vehicular Communications in Drive-thru Internet

Drive-thru Internet can provide vehicles intermittent wireless connections to the Internet when driving through the coverage areas of RSUs, and accordingly enjoy the Internet services [1]. To well study the Drive-thru Internet communication framework and the related vehicular services applications on it, this chapter provides the overviews of the vehicular communications in Drive-thru Internet in terms of the characteristics of Drive-thru Internet, medium access control, and vehicular content distribution approaches. The remainder of this chapter is organized as follows. Section 2.1 studies the Drive-thru Internet characteristics and related research issues. Section 2.2 introduces the vehicular access control related literature in Drive-thru Internet. Section 2.3 surveys the vehicular content distribution related literature in Drive-thru Internet.

2.1 Drive-thru Internet Characteristics

In spite of the increasing research popularity of VANET oriented applications using WiFi as a wireless access scenario, the majority of research works focusing on wireless local area networks do not consider the specific vehicular environment. The overview of the performance evaluation and related enhancement strategies in wireless local area networks is helpful to better understand the principles of Drive-thru Internet and accordingly improve the Drive-thru Internet performances

© The Author(s) 2015
H. Zhou et al., *Cooperative Vehicular Communications in the Drive-thru Internet*,
SpringerBriefs in Electrical and Computer Engineering,
DOI 10.1007/978-3-319-20454-3_2

Table 2.1 Capacity, coverage, and deployment related data of WLAN

802.11 protocol	Data rate (Mbps)	Throughput (Mbps)	Channel number	Capacity (Mbps)	Coverage (m)
802.11b	11	6	3	18	248
802.11g	54	22	3	66	248
802.11a	54	25	12	300	100

for vehicular applications. Therefore, we first present the Drive-thru Internet characteristics from the following four aspects and then provide the related research issues:

- Intermittent connectivity. Seen from Table 2.1,[1] Cisco white paper in [2] reports the capacity, coverage, and deployment related data of WLAN. Even though the WLANs have high throughput and network capacity, the coverage of typical IEEE 802.11 network is short, and the data in Table 2.1 shows that the largest coverage range of WLANs in open indoor office environment is no more than 250 m. Hence, considering the fast driving feature of vehicular mobility and widely but unevenly deployed WiFi hotspots, the vehicle-to-RSU connectivity emerges with intermittence feature, which is the main limitation of Drive-thru Internet for vehicular applications. To cope with the connectivity feature of Drive-thru Internet, references [3–6] investigate how to provide the application performance guarantees for vehicular users in Drive-thru Internet.
- Performance anomaly. Due to the multi-rate feature of 802.11 WLAN, the low transmission rate will greatly reduce the average throughput over all vehicles in the RSU's coverage, which is referred as the performance anomaly. The performance anomaly phenomenon was first reported in [7], which is taken IEEE 802.11b networks for instance. The performance anomaly of Drive-thru Internet can lead the fairness and effectiveness issues of vehicular access. In term of IEEE 802.11 DCF related performance evaluation and enhancement in general wireless networks, the majority of literature investigated the performance anomaly impact on the IEEE multi-rate 802.11 networks and corresponding improving strategies. In [8–11], the analytical performance anomaly in IEEE 802.11 networks with multiple transmission rates is demonstrated and some possible remedies to improve the performance in IEEE 802.11 networks are developed, such as optimally selecting the initial contention window, frame size, maximal backoff stage, etc., depending on the PHY-layer transmission rates and

---

[1]The communication parameter settings for the measurement of IEEE 802.11a/b/g networks in Table 2.1 are 40 mW with 6 dBi, 100 mW with 2.2 dBi, and 30 mW with 2.2 dBi gain diversity patch antenna, respectively, in the open indoor office environment. Even though the measurement data is just for open indoor office environment, this is only available complete test data, and compared with the investigated 802.11b networks model in [1], there is few difference for the test data. In addition, this test data of IEEE 802.11g is based on the setting considerations of no legacy support and no 802.11b clients in cell.

traffic states. In [12], how to overcome the performance anomaly in Drive-thru Internet is studied and accordingly a fair and effective spatial medium sharing approach is proposed.

- Limited transmitted data. Due to the high mobility, harsh and intermittent wireless channels, the transmitted data volume of individual vehicle per drive-thru is quite limited as observed in real-world tests. To achieve the integrity and high-throughput featured vehicular content distribution applications, e.g., video and online streaming, the cooperative vehicular communications is needed in Drive-thru Internet. To enhance the transmitted data volume size for meeting the quality of service (QoS) requirement of vehicular applications in Drive-thru Internet, in [13–18], the cooperative download and relay schemes are proposed.
- Frequent handover. On the one hand, the frequent handover and association switch among different WiFi hotspots are because of the intermittent connectivity and medium contending feature of CSMA/CA based vehicular access control protocols. The frequent handover will highly impact on the reliability and robustness of vehicular applications in Drive-thru Internet, especially for many delay and throughput sensitive applications. Therefore the fair and effective association schemes in Drive-thru Internet are required. On the other hand, for some mobile offloading applications, due to the frequent handover features of vehicular access, it is challenging to achieve the mobile offloading optimization objectives, which are always tradeoff issues, e.g., WiFi handover frequency vs. access cost; nodal throughput optimization vs. access fairness. To address those issues, in [19–21], how to achieve the efficiency and fairness for association control in Drive-thru Internet is investigated.

2.2 Vehicular Access Control

Different from general IEEE 802.11 networks, the introduced vehicular environment makes Drive-thru Internet unique in terms of the traffic features. In [22], the impact of vehicular mobility on the MAC of Drive-Thru Internet is investigated, and the corresponding MAC parameters adjustment in tune with the nodal mobility for throughput improvement of 802.11 DCF is proposed. In [23], the impact of vehicular mobility on IEEE 802.11 p infrastructureless vehicular networks is investigated. In [24], the impact of road traffic on the throughput performance of Drive-thru Internet is discussed, and a MAC protocol to control the transmission probability of vehicles for maximizing system throughput is proposed. In [25], the uplink MAC performance of Drive-thru Internet is evaluated. It is worthy of mention that this chapter points out that the higher throughput can be achieved by determining the optimal admission control region by using simulations. However, how to reach the optimal drive-thru throughput for different traffic flow states and multi-rate 802.11 network models is lacked. In addition, similar with the cooperative relay approach proposed in [26], in which nodes located within high transmission rate region will repeat/relay the data to the node with low transmission rate to address the rate anomaly problem. In [27], a multi-vehicular maximum (MV-MAX) approach is introduced. However, those relay based approaches are less fair than the IEEE 802.11 DCF.

2.3   Vehicular Content Distribution

Cooperative vehicular communications represent an effective approach to extend the RSU's coverage and have attracted an extensive research attention in the past. Various cooperative schemes have been proposed in VANET accordingly, which can be usually divided into two categories [28], i.e., vehicle-to-RSU (V2R) and V2V communications. In the following, we present the up-to-date cooperative vehicular content distribution literature overview for both infrastructure supportive content distribution research and V2V relay supportive content distribution research.

2.3.1   Infrastructure Supportive Content Distribution

For cooperative download of large-sized content from roadside infrastructure in V2R communications, Trullols-Cruces et al. [29] considers an urban/suburban scenario, unlike the highways scenario, which is hard to predict the contacts between cars. Trullols-Cruces et al. [29] devises the solutions for selecting the right vehicles to carry data chunks and assigning the data chunks among the selected vehicles. To improve the diversity of information circulating in vehicular networks, Zhang and Yeo [30] proposes a cooperative content distribution system to distribute contents to moving vehicles via APs' collaboration. Based on the vehicular contact patterns observed by APs, the shared content can be prefetched in the selected representative APs, and the vehicles can obtain the completed data from those selected APs. Saad et al. [31] investigates the cooperative strategy among the RSUs by applying a coalition formation game, which coordinates the classes of data to serve their responsible vehicles. In addition, Liang and Zhuang [16] investigates the utilization of roadside wireless local area networks as a network infrastructure for data dissemination, where the network-level and packet-level cooperation are both exploited.

2.3.2   V2V Relay Supportive Content Distribution

In terms of cooperative V2V communications, Nandan et al. [32] proposes a swarming protocol and a piece-selection strategy to cooperatively exchange the pieces for the same content sharing among different vehicles. Sardari et al. [33] investigates a V2V collaboration scenario for distributing data to sparse vehicular networks, and by applying the rateless coding approach and using vehicles as data carriers, the reliable dissemination performance can be achieved. Trullols-Cruces et al. [34] addresses V2V cooperation issue by introducing a delay cooperative ARQ mechanism and vehicular routes predictability based carry-and-forward mechanism. To combat the lossy wireless transmissions and achieve high speed content sharing, the symbol level network coding is adopted in V2V communications. Yan et al. [35] analyzes the throughput of cooperative content distribution from RSUs. Due to the observation that the vehicles which are nears from the AP have high signal strength, Zhao et al. [36] proposes a relay-based solution to extend the service range of roadside APs. In addition, Li et al. [37] designs a push-based popular content distribution scheme where contents are actively broadcasted to vehicles from RSUs and further distributed cooperatively among vehicles. Wang et al. [38] discusses the maximum achievable amount of information that can be relayed forwarding along a vehicular stream. Brandner et al. [39] evaluates the packet delivery performance of low complex cooperative relaying by real-world tests. Wang et al. [40] introduces a coalitional graph game to model the cooperative vehicles for popular content distribution.

In addition, the high mobility, intermittent connectivity, and unreliability of the wireless channel can affect the cooperative performance of VANET. Considering the aforementioned factors, to satisfy the need for massive data download and forwarding applications, [41, 42] both investigate the cooperative MAC protocols in vehicular networks.

2.4 Summary

This chapter has introduced the Drive-thru Internet characteristics and surveyed the existing literature for cooperation issue in Drive-thru Internet in terms of the medium access control. It has also presented a comprehensive overview of vehicular content distribution approaches for a better understanding of the related research.

References

1. W.L. Tan, W.C. Lau, O. Yue, T.H. Hui, Analytical models and performance evaluation of drive-thru internet systems. IEEE J. Sel. Areas Commun. **29**(1), 207–222 (2011)
2. R. Seide, Capacity, coverage, and deployment considerations for ieee 802.11 g. Cisco Systems White Paper, San Jose, CA (2005)
3. Z. Zheng, P. Sinha, S. Kumar, Alpha coverage: bounding the interconnection gap for vehicular internet access, in *Proceedings of IEEE INFOCOM* (IEEE, Rio De Janeiro, 2009), Rio De Janeiro, Brazil, pp. 2831–2835
4. M.J. Khabbaz, H.M. Alazemi, C.M. Assi, Delay-aware data delivery in vehicular intermittently connected networks. IEEE Trans. Commun. **61**(3), 1134–1143 (2013)
5. S.F. Hasan, X. Ding, N.H. Siddique, S. Chakraborty, Measuring disruption in vehicular communications. IEEE Trans. Veh. Technol. **60**(1), 148–159 (2011)

6. Z. Zheng, Z. Lu, P. Sinha, S. Kumar, Maximizing the contact opportunity for vehicular internet access, in *Proceedings of IEEE INFOCOM* (IEEE, San Diego, 2010), San Diego, CA, pp. 1–9
7. M. Heusse, F. Rousseau, G. Berger-Sabbatel, A. Duda, Performance anomaly of 802.11 b, in *Proceedings of INFOCOM*, vol. 2 (IEEE, San Francisco, 2003), San Francisco, pp. 836–843
8. D.B. Rawat, D.C. Popescu, G. Yan, S. Olariu, Enhancing vanet performance by joint adaptation of transmission power and contention window size. IEEE Trans. Parallel Distrib. Syst. **22**(9), 1528–1535 (2011)
9. D.-Y. Yang, T.-J. Lee, K. Jang, J.-B. Chang, S. Choi, Performance enhancement of multirate ieee 802.11 wlans with geographically scattered stations. IEEE Trans. Mob. Comput. **5**(7), 906–919 (2006)
10. Y.-L. Kuo, K.-W. Lai, F.-S. Lin, Y.-F. Wen, E.-K. Wu, G.-H. Chen, Multirate throughput optimization with fairness constraints in wireless local area networks. IEEE Trans. Veh. Technol. **58**(5), 2417–2425 (2009)
11. P. Lin, W.-I. Chou, T. Lin, Achieving airtime fairness of delay-sensitive applications in multirate ieee 802.11 wireless lans. IEEE Commun. Mag. **49**(9), 169–175 (2011)
12. H. Zhou, B. Liu, F. Hou, T.H. Luan, N. Zhang, L. Gui, Q. Yu, X. Shen, Spatial coordinated medium sharing: optimal access control management in drive-thru internet. IEEE Trans. Intell. Transp. Syst. 1–14 (2015)
13. J. Yoo, B.S.C. Choi, M. Gerla, An opportunistic relay protocol for vehicular road-side access with fading channels, in *Proceedings of IEEE ICNP* (IEEE, Kyoto, 2010), Kyoto, Japan, pp. 233–242
14. H. Zhou, B. Liu, T.H. Luan, F. Hou, L. Gui, Y. Li, Q. Yu, X. Shen, Chaincluster: engineering a cooperative content distribution framework for highway vehicular communications. IEEE Trans. Intell. Transp. Syst. **15**(6), 2644–2657 (2014)
15. N. Kumar, J.J. Rodrigues, N. Chilamkurti, Bayesian coalition game as a service for content distribution in internet of vehicles. IEEE Internet Things J. **1**(6), 544–555 (2014)
16. H. Liang, W. Zhuang, Cooperative data dissemination via roadside wlans. IEEE Commun. Mag. **50**(4), 68–74 (2012)
17. K. Ota, M. Dong, S. Chang, H. Zhu, Mmcd: cooperative downloading for highway vanets. IEEE Trans. Emerg. Top. Comput. **3**(1), 34–43 (2015)
18. M. Wang, Q. Shen, R. Zhang, H. Liang, S. Shen, Vehicle-density based adaptive mac for high throughput in drive-thru networks. IEEE Internet Things J. **1**(6), 533–543 (2014)
19. L. Xie, Q. Li, W. Mao, J. Wu, D. Chen, Association control for vehicular wifi access: pursuing efficiency and fairness. IEEE Trans. Parallel Distrib. Syst. **22**(8), 1323–1331 (2011)
20. Y.-S. Chen, M.-C. Chuang, C.-K. Chen, Deucescan: deuce-based fast handoff scheme in ieee 802.11 wireless networks. IEEE Trans. Veh. Technol. **57**(2), 1126–1141 (2008)
21. K. Shafiee, A. Attar, V.C. Leung, Optimal distributed vertical handoff strategies in vehicular heterogeneous networks. IEEE J. Sel. Areas Commun. **29**(3), 534–544 (2011)
22. T.H. Luan, X. Ling, X. Shen, Mac in motion: impact of mobility on the mac of drive-thru internet. IEEE Trans. Mob. Comput. **11**(2), 305–319 (2012)
23. W. Alasmary, W. Zhuang, Mobility impact in ieee 802.11 p infrastructureless vehicular networks. Ad Hoc Netw. **10**(2), 222–230 (2012)
24. K. Kim, J. Lee, W. Lee, A mac protocol using road traffic estimation for infrastructure-to-vehicle communications on highways. IEEE Trans. Intell. Transp. Syst. **14**(3), 1500–1509 (2013)
25. Y. Zhuang, J. Pan, V. Viswanathan, L. Cai, On the uplink Mac performance of a drive-thru internet. IEEE Trans. Veh. Technol. **61**(4), 1925–1935 (2012)
26. P. Bahl, R. Chandra, P.P. Lee, V. Misra, J. Padhye, D. Rubenstein, Y. Yu, Opportunistic use of client repeaters to improve performance of wlans. IEEE/ACM Trans. Netw. **17**(4), 1160–1171 (2009)
27. D. Hadaller, S. Keshav, T. Brecht, Mv-max: improving wireless infrastructure access for multi-vehicular communication, in *Proceedings of the SIGCOMM Workshop on Challenged Networks* (ACM, Pisa, 2006), Pisa, Italy, pp. 269–276

28. P. Alexander, D. Haley, A. Grant, Cooperative intelligent transport systems: 5.9-ghz field trials. Proc. IEEE **99**(7), 1213–1235 (2011)
29. O. Trullols-Cruces, M. Fiore, J. Barcelo-Ordinas, Cooperative download in vehicular environments. IEEE Trans. Mob. Comput. **11**(4), 663–678 (2012)
30. D. Zhang, C. Yeo, Enabling efficient wifi-based vehicular content distribution. IEEE Trans. Parallel Distrib. Syst. **24**(3), 233–247 (2013)
31. W. Saad, Z. Han, A. Hjorungnes, D. Niyato, E. Hossain, Coalition formation games for distributed cooperation among roadside units in vehicular networks. IEEE J. Sel. Areas Commun. **29**(1), 48–60 (2011)
32. A. Nandan, S. Das, G. Pau, M. Gerla, M. Sanadidi, Cooperative downloading in vehicular ad-hoc wireless networks, in *Proceedings of WONS* (2005), pp. 32–41
33. M. Sardari, F. Hendessi, F. Fekri, Infocast: a new paradigm for collaborative content distribution from roadside units to vehicular networks, in *Proceedings of IEEE SECON* (2009), pp. 1–9
34. O. Trullols-Cruces, J. Morillo-Pozo, J.M. Barcelo, J. Garcia-Vidal, A cooperative vehicular network framework, in *Proceedings of IEEE ICC* (2009), pp. 1–6
35. Q. Yan, M. Li, Z. Yang, W. Lou, H. Zhai, Throughput analysis of cooperative mobile content distribution in vehicular network using symbol level network coding. IEEE J. Sel. Areas Commun. **30**(2), 484–492 (2012)
36. J. Zhao, T. Arnold, Y. Zhang, G. Cao, Extending drive-thru data access by vehicle-to-vehicle relay, in *Proceedings of ACM VANET* (2008), pp. 66–75
37. M. Li, Z. Yang, W. Lou, Codeon: cooperative popular content distribution for vehicular networks using symbol level network coding. IEEE J. Sel. Areas Commun. **29**(1), 223–235 (2011)
38. Q. Wang, P. Fan, K.B. Letaief, On the joint v2i and v2v scheduling for cooperative vanets with network coding. IEEE Trans. Veh. Technol. **61**(1), 62–73 (2012)
39. G. Brandner, U. Schilcher, T. Andre, C. Bettstetter, Packet delivery performance of simple cooperative relaying in real-world car-to-car communications. IEEE Wirel. Commun. Lett. **1**(3), 237–240 (2012)
40. T. Wang, L. Song, Z. Han, Coalitional graph games for popular content distribution in cognitive radio vanets. IEEE Trans. Veh. Technol. **60**(99), 1–10 (2013)
41. J. Zhang, Q. Zhang, W. Jia, Vc-mac: a cooperative mac protocol in vehicular networks. IEEE Trans. Veh. Technol. **58**(3), 1561–1571 (2009)
42. S. Bharati, W. Zhuang, Cah-mac: cooperative adhoc mac for vehicular networks. IEEE J. Sel. Areas Commun. **31**(9), 470–479 (2013)

Chapter 3
Spatial Coordinated Medium Sharing in the Drive-thru Internet

The advance of wireless communications and pervasive use of mobile electronics in the recent years have driven the ever-increasing user demands of ubiquitous Internet access [1]. This is particularly evident for in-vehicle Internet access with people now spending much their time in cars [2]. As a result, an extensive body of research has been devoted to enabling vehicular communications with diverse applications ranging from the road safety, trip entertainment to driving efficiency and traffic management [3, 4].

In this chapter, we propose a unified analytical framework for Drive-Thru Internet performance evaluation, which can be applied to the widely deployed multi-rate IEEE 802.11 WLANs and adapt to different traffic flow states scaled from the free flow state to congested flow state. Towards this end, we first apply the fluid traffic motion (FTM) model which is widely used in the traffic engineering analysis and empirical data collected from the traffic flow report [5] to analyze the contending vehicle number and residence time within distinguished length of spatial zones within AP's coverage mathematically. By accurately modeling and formulating the relationship between the coverage range and data rate, which is abstracted from the real-world measurement data of IEEE 802.11 WLANs, we present the mathematical expression of mean vehicular saturated throughput and transmitted data volume per drive-thru in IEEE 802.11 networks with DCF. Particularly, observed from the unique mobility feature of vehicles in VANET, i.e., the relatively fair medium access opportunity during the AP's sojourn time, almost similar velocity, and the same mobility direction on each driving lane, we can schedule the vehicles to transmit in the optimally selected spatial region of coverage range, which can achieve maximal mean saturated throughput and transmitted data volume, which are validated using numerical analysis.

© The Author(s) 2015
H. Zhou et al., *Cooperative Vehicular Communications in the Drive-thru Internet*,
SpringerBriefs in Electrical and Computer Engineering,
DOI 10.1007/978-3-319-20454-3_3

The remainder of this chapter is organized as follows. Section 3.1 describes the network model. Section 3.2 presents the problem formulation. Section 3.3 designs the optimal access control management scheme. Section 3.4 validates the analysis accuracy with simulations, and Sect. 3.5 closes the chapter with conclusions.

3.1 System Model

We consider the Drive-thru Internet to support various vehicle-oriented applications, such as multimedia content distribution, file downloading/uploading, and web browsing, etc., shown in Fig. 3.1. An RSU, i.e., AP, is installed on the roadside which serves as the gateway to provide Internet services to vehicles driving through the AP's coverage. As discussed in [6, 7], the multi-rate spatial zones within AP coverage can be divided into three types according to the achieved transmission throughput when driving through: (1) entry zone; (2) production zone; (3) exit zone. Within distinct zones, the achieved transmission bit-rate by connected vehicles will be totally different.

However, the well-known performance anomaly problem is existed in the Drive-thru Internet and degrades its performance as well. Taken IEEE 802.11b networks for example, and shown in Fig. 1.3. We assume a number of vehicles contend the medium using carrier sense multiple access with collision avoidance (CSMA/CA) synchronously. Because the transmission rates from AP to the vehicles within the coverage is 11, 5.5, 2, 1 Mb/s, respectively, depending on the vehicular distance from AP. According to the measurement data in Table 3.1 and shown in Fig. 3.2, if a vehicle locates within range of 270 and 410 ft and has the current frame

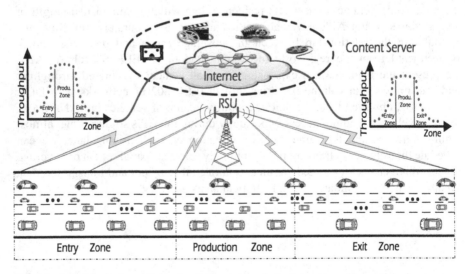

Fig. 3.1 A widely deployed Drive-thru Internet scenario for Internet service provisioning

Table 3.1 Measurement data in an open indoor office environment for 802.11a, 802.11b, and 802.11g (in unit of ft)

	Date rate (Mbps)											
Protocol	54	48	36	24	18	12	11	9	6	5.5	2	1
IEEE802.11a	45	50	65	85	110	130	–	150	165	–	–	–
IEEE802.11b	–	–	–	–	–	–	160	–	–	220	270	410
IEEE802.11g	90	95	100	140	180	210	160	250	300	220	270	410

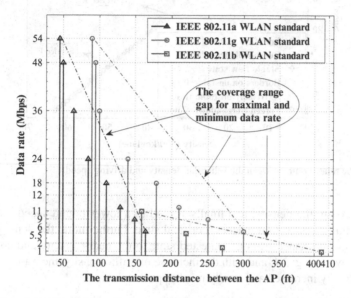

Fig. 3.2 The relationship between the transmitted data rate and the coverage range of AP

transmission opportunity, the transmission link rate can be predicted as 1 Mbs. Definitely, using low transmission rate, the time duration to transmit a frame by AP will be longer than that from a vehicle who transmits using higher transmission rate, which therefore will greatly reduce the mean individual throughput of the vehicles within the AP's coverage. In this chapter, by considering different traffic flow states and combining the empirical data collected from the real IEEE 802.11 networks test, to achieve best mean drive-thru throughput for all the vehicles within certain region of AP's coverage, the vehicles will be scheduled for optimal spatial coordinated medium sharing in a distributed and fair way.

3.1.1 Macroscopic Vehicular Mobility

We apply the FTM model to capture the macroscopic relationship between the average vehicular speed and traffic density on the road [5, 8], which can be expressed in (3.1). As shown in Fig. 3.3, the vehicular speed is monotonically decreased with

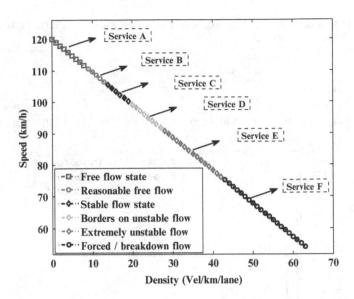

Fig. 3.3 The relationship between the vehicular density and driving speed

the increased traffic density, and finally forced to a lower bound speed, i.e., v_{min}, when the traffic congestion reaches a critical state. Furthermore, the density-traffic flow relationship scaled from uncontested state to the near-capacity and congested state is shown in Fig. 3.4, and with the increase of traffic density, the traffic flow can be dramatically increased and finally reach a peak value.

$$\bar{v} = \max\{v_{min}, v_{max}(1 - \frac{\sigma}{\sigma_{jam}})\} \tag{3.1}$$

where $\bar{v}$, v_{min}, and v_{max} denote the vehicles' average, minimum, and maximal speed, respectively. σ and σ_{jam} are the factual traffic density and jammed traffic density, respectively. To analyze the whole vehicular traffic flow process including the blocked traffic flow state, we set $v_{min} = 0$ in our analysis and simulation. σ_{jam} is determined by the capacity of roadway segment, and with the unit of vehicular number per meter.

As specified in Table 3.2, according to different traffic flow states [5, 9], there are seven-level traffic services, which cover the traffic flow states ranging from the uncontested traffic state, near capacity traffic state to the congested traffic state. Considering a heavy congestion case on the highways, the traffic flow basically cannot move, hence, it is meaningless for the vehicular drive-thru throughput analysis, so here we only consider the flow operation scenarios of first six levels of traffic services, which are investigated in [5]. For a roadway segment with a length L, the mean vehicle residence time and mean number of vehicles within the L roadway segment are denoted by $\overline{\Gamma}$ and $\overline{N}$, respectively, where $\overline{\Gamma} = L/\bar{v}$ and

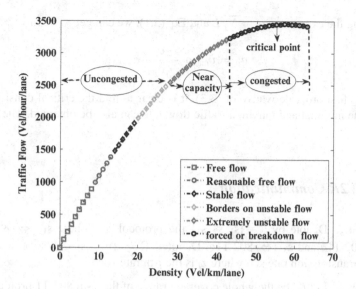

Fig. 3.4 The relationship between the vehicular density and traffic flow

Table 3.2 Traffic flow state per lane for different densities

Density (veh/km)	Traffic flow state		
	TSL	Speed (km/h)	Flow operations
0–8	A	≥ 97	Free
9–13	B	≥ 92	Reasonable free
14–19	C	≥ 87	Stable
20–27	D	≥ 74	Borders on unstable
28–42	E	≥ 49	Extremely unstable
43–63	F	< 49	Forced or breakdown

$0 \leq \overline{N} \leq L \cdot \sigma_{jam}$. Denoted by ℓ the mean length of vehicles, σ_{jam} can be calculated as $\sigma_{jam} = 1/\ell$. Generally, the mean length of vehicles ℓ is about 6 m. To consider the driving space between the vehicles for safety issue, the typical value of σ_{jam} is set as $115/1000$ [5]. For a mean vehicle arrival rate ϑ_v, the relationship between ϑ_v and σ can be established in Lemma 1:

Lemma 1. *The critical density value σ_c for vehicle arrival rate ϑ_v is $\sigma_{jam}/2$, and ϑ_v is directly increasing with σ when $\sigma \in [0, (\sigma_{jam}/2)]$ and inverse proportion when $\sigma \in (\sigma_{jam}/2)$. The range of traffic flow rate is $[0, \upsilon_{max}\sigma_{jam}/4]$.*

Proof. Using the Little's law, we can get that the mean vehicle arrival rate can be formulated as

$$\vartheta_v = \bar{N}/\bar{\Gamma} = \bar{N}/(L/\bar{\upsilon}).   \tag{3.2}$$

Combining the expression $\sigma = \overline{N}/L$ and Eq. (3.1), we can get

$$\vartheta_v = \upsilon_{max}\sigma - \frac{\upsilon_{max}}{\sigma_{jam}}\sigma^2, \tag{3.3}$$

By setting first order derivative of (3.3), it is clear to find the critical density value σ_c, and the minimal and maximal traffic flow rate can also be obtained. The lemma is proved. ∎

3.1.2 V2R Communication

Definition 1. Denote by $\mathbb{S} \overset{def}{=} \{s_1, s_2, s_3\}$ the protocol set, where $s_1 : \Leftrightarrow$ 802.11b, $s_2 : \Leftrightarrow$ 802.11a and $s_3 : \Leftrightarrow$ 802.11g. Denoted $\mathbb{C} \triangleq \{v_{s_i,z_j}|s_i \in \mathbb{S}, z_j \in \mathbb{Z}\}$ be the adaptive transmission rate set, where z_j is the j-th rate zone.

Definition 2. Let L_{s_i} be the whole coverage range of the s_i-th 802.11 protocol, and denote by $\mathscr{R}$ be the coverage set of different transmission rates, where $r_{s_i,z_j} \triangleq |x_{s_i,z_{j+1}} - x_{s_i,z_j}|$ is the coverage range of j-th transmission rate zone, x_{s_i,z_j} is the coverage range point of j-mode transmission rate zone using s_i-th 802.11 protocol, and $L_{s_i} = \sum_{z_j \in \mathbb{Z}} r_{s_i,z_j}$.

$$\mathscr{R} = \begin{bmatrix} r_{s_1,z_1} & r_{s_1,z_2} & \cdots & r_{s_1,z_j} & \cdots & r_{s_1,z_{12}} \\ r_{s_2,z_1} & r_{s_2,z_2} & \cdots & r_{s_2,z_j} & \cdots & r_{s_2,z_{12}} \\ r_{s_3,z_1} & r_{s_3,z_2} & \cdots & r_{s_3,z_j} & \cdots & r_{s_3,z_{12}} \end{bmatrix}$$

$$= \begin{bmatrix} 45 & 5 & 15 & 20 & 25 & 20 & - & 20 & 15 & - & - & - \\ - & - & - & - & - & - & 160 & - & - & 60 & 50 & 140 \\ 90 & 5 & 5 & 40 & 20 & 20 & 30 & 10 & 30 & 20 & 30 & 110 \end{bmatrix} \tag{3.4}$$

We consider a unified multi-rate IEEE 802.11 transmission rate model, which is denoted in Definitions 1 and 2. In specific, for any vehicle n within RSU's coverage, $n \in \mathcal{N}$, the transmission data rate $v_{s_i}^n$ from AP to the vehicle n when using s_i-th protocol is closely related to the channel fading and interference, which can be expressed as a function of distance from RSU to the receiving vehicle, and the transmission rates in z_j-zone are expressed as (3.5).

$$v_{s_i}^n(t) = \{v_{s_i,z_j}|x_{s_i,z_j} \leq \bar{\upsilon} \cdot \Gamma(t) < x_{s_i,z_{j+1}}\}, n \in \mathcal{N}, \tag{3.5}$$

where $\Gamma(t)$ is the sojourn time within RSU's coverage using mean driving speed $\bar{\upsilon}$.

Table 3.1 provides the measurement data of three types of IEEE 802.11 protocols under different communication setting [10].[1] We show the relationship between the distance d_n and transmission rate in Fig. 3.2, in which the unit of distance and transmission rate is with ft and Mbps, respectively. Clearly, except in few special distance points in Fig. 3.2, i.e., 160, 220, and 270 ft for 802.11g networks, the transmission rate using all the 802.11 protocols in $\mathbb{S}$ is in vice direct proportion to the coverage range.

3.2 Problem Formulation

We target to analyze the vehicular drive-thru throughput and the total download volume applying the unified multi-rate transmission rate model in IEEE 802.11 series protocols. To analyze the vehicular drive-thru throughput performance, we first analyze the mean contending vehicular number and residence time within AP's coverage. Then, we analyze the drive-thru throughput of multi-rate IEEE 802.11 networks with DCF. Finally, we formulate the optimal time coordinated channel access region in multi-rate WLANs to optimize the vehicular drive-thru throughput and whole transmitted data volume.

3.2.1 Contending Vehicular Number and Residence Time

Note that the coverage range of AP along the road is symmetric, and let L_{s_i} be the first half (mirror) coverage range of the s_i-th 802.11 protocol and r_{s_i,z_j} is the first half (mirror) coverage range of j-th transmission rate zone. For simplification, we neglect the notion s_i in the following rest space, i.e., we take one type of 802.11 protocol for example, which can be extended to the unified multi-rate transmission rate analytical model. According to investigation in [5], the speed v_i of arbitrary arriving vehicle i follows the pdf of normal distribution, which is given by

$$f_v(v_i) = \frac{1}{\delta_v \sqrt{2\pi}} e^{-\left(\frac{v_i - \bar{v}}{\delta_v \sqrt{2}}\right)^2},$$ (3.6)

---

[1]The current communication settings in the measurement of 802.11a/b/g networks are 40 mW with 6 dBi, 100 mW with 2.2 dBi, and 30 mW with 2.2 dBi gain diversity patch antenna, respectively. Even though the measurement data is from open indoor office environment, according to the investigation of 802.11b networks in [7], we believe the measurement data can be suitable to the outdoor environment as well. In addition, this is data is rational and feasible for the urban case with high buildings. Hence, we use this measurement data for the outdoor Drive-thru Internet performance analysis.

where $\bar{v} \in [v_{\min}, v_{\max}]$, $\delta_v = k\bar{v}$ and $v_{\min} = \bar{v} - m\delta_v$. As it's justified in [11], δ_v and the two-tuple (k, m) can be found in the typical values of velocity distributions, which is based on the experimental data. To avoid generating the negative speeds or some values approaching to zero, a truncated version of vehicular speed distribution will replace (3.6). By defining the function $\text{erf}(x) = \frac{2}{\pi} \int_0^x e^{-t^2} dt$, and the truncated vehicular speed distribution can be rewritten as

$$
\begin{aligned}
\widehat{f_v}(v_i) &= \frac{f_v(v_i)}{\int_{v_{\min}}^{v_{\max}} f_v(v_i) dv_i} \\
&= \frac{f_v(v_i)}{\frac{1}{2} \text{erf}(\frac{v_{\max} - \bar{v}}{\delta_v \sqrt{2}}) - \frac{1}{2} \text{erf}(\frac{v_{\min} - \bar{v}}{\delta_v \sqrt{2}})}.
\end{aligned}
\tag{3.7}
$$

Lemma 2. *Given the coverage range L and the speed of arbitrary arriving vehicle follows the normal distribution, the PDF of vehicle's residence time $f_\Gamma^L(\tau)$ is as follows,*

$$
f_\Gamma^L(\tau) = \frac{M \cdot L}{\tau^2 \delta_v \sqrt{2\pi}} e^{-(\frac{\frac{L}{\tau} - \bar{v}}{\delta_v \sqrt{2}})^2}, \tau \in [\frac{L}{v_{\max}}, \frac{L}{v_{\min}}]
\tag{3.8}
$$

Proof. The real-world road test indicates that the vehicle's speed can keep constantly for several hundred meters. Based on the reference [11], the cumulative distribution function (CDF) of vehicle's residence time $F_\Gamma^L(\tau)$ can be derived by combining the CDF of vehicle's speed $\widehat{F_v}$, shown as

$$
F_\Gamma^L(\tau) = 1 - \widehat{F_v}(\frac{L}{\tau}) = 1 - \frac{M}{2}[1 + \text{erf}(\frac{\frac{L}{\tau} - \bar{v}}{\delta_v \sqrt{2}})]
\tag{3.9}
$$

where $M = 2 / (\text{erf}(\frac{v_{\max} - \bar{v}}{\delta_v \sqrt{2}}) - \text{erf}(\frac{v_{\min} - \bar{v}}{\delta_v \sqrt{2}}))$. By setting the derivative of the CDF of vehicle's residence time $F_\Gamma^L(\tau)$ and combining the expression $\frac{d}{dx} \text{erf}(x) = \frac{2}{\sqrt{\pi}} e^{-x^2}$, we can prove the lemma.   ∎

Based on the PDF of vehicle's residence time $f_\Gamma^L(\tau)$, we can easily get the expression of mean vehicle's residence time within a fixed distance with a length L_{s_i}, shown as

$$
\begin{aligned}
\bar{\Gamma}^L &= \int_{L/v_{\max}}^{L/v_{\min}} \tau f_\Gamma^L(\tau) d\tau \\
&= \int_{L/v_{\max}}^{L/v_{\min}} \frac{M \cdot L}{\tau \delta_v \sqrt{2\pi}} e^{-(\frac{\frac{L}{\tau} - \bar{v}}{\delta_v \sqrt{2}})^2} d\tau
\end{aligned}
\tag{3.10}
$$

Considering that multi-rate transmission zones have different individual coverage, the vehicle's residence time $\bar{\Gamma}^{r_{z_j}}$ within r_{z_j}-length road segment is

$$\bar{\Gamma}^{r_{z_j}} = \bar{\Gamma}^L \frac{r_{z_j}}{\sum_{z_j \in \mathbb{Z}} r_{z_j}}, \quad z_j \in \mathbb{Z} \tag{3.11}$$

where $r_{z_j} / \sum_{z_j \in \mathbb{Z}} r_{z_j}$ is the limiting probability that one vehicle locates within the zone z_j.

Theorem 1. *The mean number of vehicles $\overline{N}^{r_{z_j}}$ within r_{z_j}-long coverage range zone follows the Poisson Distribution, and the mean number of vehicles $\overline{N}^{r_{z_j}}$ equals to*

$$\overline{N}^{r_{z_j}} = \frac{\bar{\Gamma}^L \cdot r_{z_j} \cdot \vartheta_v}{\sum_{z_j \in \mathbb{Z}} r_{z_j}} \tag{3.12}$$

Proof. According to the proof in [11], the probability of having n vehicles within a L-length coverage range zone is

$$P_n^L = \frac{(\vartheta_v \bar{\Gamma}^L)^n e^{-\vartheta_v \bar{\Gamma}^L}}{n!} \tag{3.13}$$

Hence, we can get the mean number of vehicles $\bar{N}^L$ within an L-length coverage range zone, shown as

$$\begin{aligned}
\bar{N}^L &= \sum_{n=0}^{\infty} n \cdot P_n^L = \sum_{n=1}^{\infty} n \cdot P_n^L \\
&= \sum_{n=1}^{\infty} \frac{(\vartheta_v \bar{\Gamma}^L)^{n-1} (\vartheta_v \bar{\Gamma}^L) e^{-\vartheta_v \bar{\Gamma}^L}}{(n-1)!}
\end{aligned} \tag{3.14}$$

By applying the Taylor Expansion, i.e., $e^m = \sum_{n=0}^{\infty} \frac{(m)^n}{n!}, \forall m$, we can get that $\sum_{n=1}^{\infty} \frac{(\vartheta_v \bar{\Gamma}^L)^{n-1}}{(n-1)!} = \sum_{m=0}^{\infty} \frac{(\vartheta_v \bar{\Gamma}^L)^m}{(m)!} = e^{\vartheta_v \bar{\Gamma}^L}$. Hence, mathematically, we have the expression that the number of vehicles in L-length coverage zone, i.e., $\bar{N}^L = \bar{\Gamma}^L \cdot \vartheta_v$. Combining the limiting probability $r_{z_j} / \sum_{z_j \in \mathbb{Z}} r_{z_j}$ that the number of vehicles in the r_{z_j}-length transmission zone, we can prove the theorem. ∎

3.2.2 Vehicular Drive-thru Throughput

We consider a fundamental random medium access control scheme in IEEE 802.11 protocol [12], namely, collision avoidance carrier sense multiple access (CSMA/CA) based distrusted coordination function (DCF), to analyze the vehicular drive-thru throughput in Drive-thru Internet. In addition, to eliminate the hidden terminal problem, we consider the IEEE 802.11 standard for WLAN medium access control in which the packets are transmitted by means of RTS/CTS mechanism. For generic analysis, the maximum backoff stage denoted by m and the contention window size CW are considered. We assume that the conditional probability ξ that any vehicle transmits in the randomly chosen slot-time is no relation with the vehicles' position. Under a constant and independent conditional collision probability p, ξ can be expressed as

$$\xi = \frac{2}{1 + CW + pCW \sum_{j=0}^{m-1} (2p)^j}, p \in [0, 1] \tag{3.15}$$

when $p = 0$, we can get that $\xi = 2/(1 + CW)$, and $p = 1$, Eq. (3.15) is written as $\xi = 2/(1 + 2^m CW)$.

Within the z_j-th transmission zone, let $p_{z_j}^{tr}$ be the probability that there is at least one transmission within this zone during the considered time-slot. Given the number of vehicles within this transmission zone as $\overline{N}^{r_{z_j}}$, and mathematically, we have the probability $p_{z_j}^{tr}$ that is

$$p_{z_j}^{tr} = 1 - (1 - \xi)^{\overline{N}^{r_{z_j}}} \tag{3.16}$$

where for any IEEE 802.11 protocol, i.e., $\forall s_i$, $\overline{N}^{r_{z_j}}$ is related to the limiting probability, i.e., $r_{z_j}/\sum_{z_i \in \mathbb{Z}} r_{z_i}$, that one vehicle locates within the zone, and can be obtained using Eq. (3.12).

For the AP's coverage with multi-rate transmission zones, if multiple continual transmission zones are considered, the probability p_k^{tr} that there is at least one transmission within k continual transmission zones set $\mathbb{Z}_k$, i.e., $\mathbb{Z}_k = \{ z_1, z_2, \ldots, z_k \}$, and $k \leq |\mathbb{Z}|$, can be expressed as

$$p_k^{tr} = 1 - \prod_{j=1}^{k} (1 - \xi)^{\overline{N}^{r_{z_j}}} \tag{3.17}$$

where for $k = |\mathbb{Z}|$, which means that in a whole L-length transmission coverage of AP, Eq. (3.17) can be formulated as

$$p_k^{tr} = 1 - \prod_{j=1}^{|\mathbb{Z}|} (1 - \xi)^{\bar{N}^{r_{z_j}}} = 1 - (1 - \xi)^{\sum_{j=1}^{|\mathbb{Z}|} \frac{\bar{N}^L r_{z_j}}{\sum_{z_i \in \mathbb{Z}} r_{z_i}}}$$
$$= 1 - (1 - \xi)^{\bar{N}^L \sum_{j=1}^{|\mathbb{Z}|} \frac{r_{z_j}}{\sum_{z_i \in \mathbb{Z}} r_{z_i}}} = 1 - (1 - \xi)^{\bar{N}^L} \tag{3.18}$$

Lemma 3. *Within the specified z_j-th transmission zone, and given the number of vehicles is $\overline{N}^{r_{z_j}}$, the probability $p_{z_j}^{su}$ that a successful transmission on the considered channel equals to*

$$p_{z_j}^{su} = \frac{\bar{N}^{r_{z_j}} \xi (1 - \xi)^{(\bar{N}^{r_{z_j}} - 1)}}{1 - (1 - \xi)^{\bar{N}^{r_{z_j}}}} \tag{3.19}$$

Proof. $p_{z_j}^{su}$ can also be understood as the probability that only exactly one vehicle transmits within the specified z_j-th transmission zone, with the condition that at least one vehicle transmits factually. Hence, mathematically, we have

$$p_{z_j}^{su} = \frac{\bar{N}^{r_{z_j}} \xi (1 - \xi)^{(\bar{N}^{r_{z_j}} - 1)}}{p_{z_j}^{tr}} = \frac{\bar{N}^{r_{z_j}} \xi (1 - \xi)^{(\bar{N}^{r_{z_j}} - 1)}}{1 - (1 - \xi)^{\bar{N}^{r_{z_j}}}} \tag{3.20}$$

Using Eq. (3.16), we can prove the lemma. ∎

Lemma 4. *Given the k continual transmission zones, i.e., $k = |\mathbb{Z}_k|$, the probability p_k^{su} that a successful transmission on the considered channel equals to*

$$p_k^{su} = \frac{(\sum_{j=1}^{k} \bar{N}^{r_{z_j}}) \xi (1 - \xi)^{(\sum_{j=1}^{k} \bar{N}^{r_{z_j}} - 1)}}{1 - (1 - \xi)^{\sum_{j=1}^{k} \bar{N}^{r_{z_j}}}}, k \leq |\mathbb{Z}| \tag{3.21}$$

Proof. Based on the Lemma 3 and combining Eq. (3.17), we have the probability p_k^{su} that

$$p_k^{su} = \frac{(\sum_{j=1}^{k} N^{r_{z_j}}) \xi (1 - \xi)^{(\sum_{j=1}^{k} \bar{N}^{r_{z_j}} - 1)}}{1 - \prod_{j=1}^{k} (1 - \xi)^{\bar{N}^{r_{z_j}}}}, k \leq |\mathbb{Z}| \tag{3.22}$$

We can easily prove the lemma. Especially, if we consider a whole L-length AP's coverage, i.e., $k = |\mathbb{Z}|$, we can have the expression of $p_{|\mathbb{Z}|}^{su}$ as $p_{|\mathbb{Z}|}^{su} = \frac{\bar{N}^L \xi (1 - \xi)^{(\bar{N}^L - 1)}}{1 - (1 - \xi)^{\bar{N}^L}}$. ∎

We consider the RTS/CTS medium access control mechanism, and the collision can only happen on RTS frames. For such a mechanism, let $E[T_{su,z_j}^{rts}]$ and $E[T_{bu,z_j}^{rts}]$ be the mean time that the channel is sensed busy within z_j-th transmission zone due to the successful transmission or a collision, respectively. Mathematically, $E[T_{su,z_j}^{rts}]$ and $E[T_{bu,z_j}^{rts}]$ can be expressed as

$$\begin{cases} E[T_{su,z_j}^{rts}] = (RTS + CTS + ACK + E[P]) / v_{z_j} \\ \qquad\qquad + 3\,SIFS + 4\sigma_{propa} + DIFS \\ E[T_{bu,z_j}^{rts}] = RTS / v_{z_j} + DIFS + \sigma_{propa} \end{cases} \tag{3.23}$$

where v_{z_j} is the link transmission rates within z_j-zone, which can be obtained from the measurement data of Table 3.1.

Easily, we can extend the derivation of mean time that a successful and collided transmission within one transmission zone to the case with k continual transmission zones. Among k continual transmission zones, let $r_{z_j} / \sum_{z_j \in \mathbb{Z}_k} r_{z_j}$ be the limiting probability that a vehicle is in the transmission zone z_j. The mean time that the channel is sensed busy within k continual transmission zones are defined as $E[T_{su,k}^{rts}]$ and $E[T_{bu,k}^{rts}]$, respectively, which can be expressed as

$$
\begin{cases}
E[T_{su,k}^{rts}] = \sum_{j=1}^{k} E[T_{su,z_j}^{rts}] \dfrac{r_{s_i,z_j}}{\sum_{z_j \in \mathbb{Z}_k} r_{s_i,z_j}}, \\[4mm]
E[T_{bu,k}^{rts}] = \sum_{j=1}^{k} E[T_{bu,z_j}^{rts}] \dfrac{r_{s_i,z_j}}{\sum_{z_j \in \mathbb{Z}_k} r_{s_i,z_j}},
\end{cases}
\qquad k \le |\mathbb{Z}| \tag{3.24}
$$

We analyze the drive-thru throughput when vehicles pass k continual transmission zones. Based on the above analysis results, we can easily get that the mean time-slot $E[T_{slot}]$ using RTS/CTS mechanism in DCF is composed of three parts:

- The mean duration of empty slot-time $E[T_k^{em}]$ within k continual transmission zones with probability of $(1 - p_k^{tr})$;
- The mean duration of successful transmission $E[T_k^{su}]$ within k continual transmission zones with probability of $p_k^{tr} \cdot p_k^{su}$;
- The mean duration $E[T_k^{bu}]$ that channel is busy within k continual transmission zones with probability of $p_k^{tr}(1 - p_k^{su})$.

Mathematically, $E[T_{slot}]$ can be formulated as

$$
E[T_{slot}] = E[T_k^{em}] + E[T_k^{su}] + E[T_k^{bu}] \tag{3.25}
$$

where we have $E[T_{em}]$, $E[T_{su}]$, and $E[T_{bu}]$, respectively,

$$
\begin{cases}
E[T_k^{em}] = (1 - p_k^{tr})\sigma_{pro} \\[2mm]
E[T_k^{su}] = p_k^{tr} p_k^{su} E[T_{su,k}^{rts}] \\[2mm]
E[T_k^{bu}] = p_k^{tr}(1 - p_k^{su}) E[T_{bu,k}^{rts}]
\end{cases}
\tag{3.26}
$$

According to the normalized 802.11 system throughput definition in [12], the nodal throughput passing k continual transmission zones can be evaluated by the ratio of mean package payload size and the mean amount of payload package transmitted in a slot time successfully. Mathematically, we have

$$
\Re_k = \frac{E[X_{pl}]}{E[T_{slot}]} = \frac{p_k^{tr} p_k^{su} E[P]}{E[T_k^{em}] + E[T_k^{su}] + E[T_k^{bu}]} \tag{3.27}
$$

In fact, the values of m and CW are hardwired in the PHY layer details [12]. Intuitively, different traffic flow states will generate different values of contending vehicular number and sojourn time within RSU's coverage, hence, the nodal throughput in (3.27) is determined by the traffic flow state if the DCF related parameters (m and CW) are fixed. According to the analysis, the more number of vehicles within a fixed road segment can lead to a higher contention overhead and less sharing capacity. The function of mean total amount of transmitted data $\mathbb{F}_k(\cdot)$ from/to AP by a vehicle is determined by the nodal throughput $\Re_k$ passing k continual transmission zones and the vehicular sojourn time $\sum_{j=1}^{k} \bar{\Gamma}^{r_{z_j}}$. Mathematically, we have $\mathbb{F}_k(\cdot)$ as

$$\mathbb{F}_k(\cdot) \overset{def}{=} \Re_k \sum_{j=1}^{k} \bar{\Gamma}^{r_{z_j}}, \; z_j \in \mathbb{Z}, k \leq |\mathbb{Z}| \tag{3.28}$$

3.2.3 Optimal Spatial Coordinated Channel Access Modeling

On the one hand, mathematically, we can treat $\mathbb{F}(\cdot)$ as a function of the distributed vehicular number $\overline{N}^{r_{z_j}}$ and the vehicular sojourn time $\overline{\Gamma}^{r_{z_j}}$ within z_j-th transmission zone, and in essence, $\overline{N}^{r_{z_j}}$ and $\overline{\Gamma}^{r_{z_j}}$ is closely related to the traffic density σ. On the other hand, from the fairly medium access sharing perspective of IEEE 802.11 protocols, with the feature of multi-rate transmission mode of IEEE 802.11 protocols in different transmission zones, all the vehicles have the same probability and time to access the AP. However, the vehicles located in low-rate transmission zones will damage the fairness of other vehicular medium access sharing opportunities within AP's coverage and reduce the total mean amount of transmitted data $\mathbb{F}(\cdot)$ from/to AP by a vehicle, which is namely the performance anomaly phenomenon. We try to regulate the vehicles to utilize the high-rate transmission zones for medium access sharing based on the traffic density σ. For different traffic densities, the traffic flow state set denoted by $\mathcal{Q}_{ser}$ includes six service levels with a low-to-high traffic density range, i.e., $\mathcal{Q}_{ser} :\Leftrightarrow \{\mathcal{Q}_A, \mathcal{Q}_B, \mathcal{Q}_C, \mathcal{Q}_D, \mathcal{Q}_E, \mathcal{Q}_F\}$, and correspondingly, the mean traffic density set of six service levels of flow state is γ_{ser}, and $\gamma_{ser} = \{\gamma_{ser}^k\}$, where $k = \{A, B, C, D, E, F\}$. The motivation of this chapter is to find an optimal channel access region for all the vehicles within AP's coverage, and the vehicles can be regulated to communicate with AP within the same channel access zones by distributed spatial coordinated channel access scheduling.

We consider the first half (mirror) coverage range of the s_i-th 802.11 protocol, i.e., L_{s_i}. Assuming that $\mathcal{X}_{s_i}$ be the length of channel access region that vehicles are regulated to communicate with AP within it. Firstly, we can transform the length of road segment $\mathcal{X}_{s_i}$ to the position of transmission zones, and if the position transformation relation satisfies the following expression (3.27), the length of road segment $\mathcal{X}_{s_i}$ will cover the first k- transmission zones.

$$k = \{n \left| \sum_{j=1}^{n-1} r_{z_j} < \mathcal{X}_{s_i} \leq \sum_{j=1}^{n} r_{z_j}, n \leq |\mathbb{Z}| \right. \} \tag{3.29}$$

Easily, the mean total amount of transmitted data $\mathbb{F}_{\mathscr{X}_{s_i}}(\cdot)$ is consisted of two parts: (1) the total amount of transmitted data $\mathbb{F}_{k-1}(\cdot)$ within the first complete $k-1$ transmission zone; (2) the transmitted data $\mathbb{F}_{z_k}^p(\cdot)$ within the part of the coverage of z_k-th transmission zone. Mathematically, we have $\mathbb{F}_{\mathscr{X}_{s_i}}(\cdot)$ as

$$\mathbb{F}_{\mathscr{X}_{s_i}}(\cdot) = \mathbb{F}_k \cdot \frac{\mathscr{X}_{s_i}}{\sum_{j=1}^{k} r_{z_j}} \tag{3.30}$$

where $\frac{\mathscr{X}_{s_i}}{\sum_{j=1}^{k} r_{z_j}}$ is the proportion that the data is transmitted within the $\mathscr{X}_{s_i}$-length coverage region.

Then, we can formulate spatial coordinated channel access scheduling by finding a optimal channel access region for all the vehicles, which can be expressed as

$$\mathscr{X}_{s_i}^*(\mathscr{Q}_{type}) \triangleq \arg \max_{\mathscr{X}_{s_i} \leq L_{s_i}} \{2 \cdot \mathbb{F}_{\mathscr{X}_{s_i}}(\cdot)\}$$

$$s.\,t. \qquad\qquad Eq.\ (3.29) \tag{3.31}$$

$$s_i \in \mathbb{R}, z_j \in \mathbb{Z}$$

$$\mathscr{Q}_{type} \in \mathscr{Q}_{ser}$$

where $\mathscr{X}_{s_i}^*(\mathscr{Q}_{type})$ is the optimal mirror channel access region for vehicular medium access under the fixed traffic flow state $\mathscr{Q}_{type}$.

3.3 Optimal Access Control Management Scheme Design

By determining the best medium access control and sharing region within AP's coverage for access control, we can significantly enhance the transmitted data volume. In fact, our proposed medium access control management approach can be easily deployed into the existing IEEE 802.11 DCF protocol without involving dirty revisions. More importantly, our proposal can provide fair use of the airtime for every vehicle as they all will drive through this region. In the real-world deployment of our proposal in Drive-thru Internet, vehicles can either locate the position within AP's coverage using the on-board GPS, or through channel measurement. If the location of a vehicle n, i.e., loc_n is not located within the selected symmetrical mirror region $[-\mathscr{X}_{s_i}^*(\mathscr{Q}_{type}), +\mathscr{X}_{s_i}^*(\mathscr{Q}_{type})]$, the vehicle n halts the transmission in a coordinated way. To avoid some dishonest nodes which continue to transmit when out of the optimal access control region, AP can stop the transmissions by using the bit rate sensing technique. The algorithm of the optimal access control management process is depicted in Algorithm 1.

Algorithm 1: Spatial coordinated medium sharing approach

Input: $\mathcal{Q}_{ty} \in \mathcal{Q}_{ser}, s_i \in \mathbb{S}, v_{max}, \sigma_{jam}$

Output: $\mathcal{X}_{s_i}^*$

1 Calculating $\overline{N}^{r_{s_i,z_j}}$ and $\overline{\Gamma}^{r_{s_i,z_j}}$;

2 Finding $\mathcal{X}_{s_i}^*(\mathcal{Q}_{type})$ i.e., $\mathcal{X}_{s_i}^*(\mathcal{Q}_{type}) \triangleq \arg\max_{\mathcal{X}_{s_i} \leq L_{s_i}} \{2 \cdot (\mathbb{F}_{k-1}(\cdot) + \mathbb{F}_{z_k}^p(\cdot))\}$; **for** $n = 1; n \leq |\mathcal{N}|$ **do**

3   **while** $loc_n \not\subset [-\mathcal{X}_{s_i}^*(\mathcal{Q}_{type}), +\mathcal{X}_{s_i}^*(\mathcal{Q}_{type})]$ **do**

4   vehicle n stops the transmission;

5   **if** *vehicle n does not stop the transmission* **then**

6   Calculating $\mathbb{C}(\mathcal{X}_{s_i}^*)$ and Sensing v_{s_i,z_j} ;

7   **while** $v_{s_i,z_j} \notin \mathbb{C}(\mathcal{X}_{s_i}^*)$ **do**

8 | AP stops the transmission of vehicle n;

9   **end while**

10   **end if**

11   **end while**

12 **end for**

3.4 Simulation Results

In this section, we evaluate the performance of our proposal using simulations. Our simulations are conducted over Matlab. Table 3.3 summarize the setting of simulation parameters, including the vehicular traffic parameters, V2R communication model parameters, and IEEE 802.11 DCF parameters, respectively.

3.4.1 Vehicular Traffic Statistics

Figure 3.5 shows the distribution of vehicle's residence time in AP when the coverage range of AP is 50 m, 100 m, and 150 m, respectively. Using our analysis shown in (10), we could evaluate the average residence time within different coverage ranges. However, due to the complexity of calculating integral in (10), we evaluate the average vehicular residence time within a fixed coverage range

Table 3.3 Setting of DCF mechanism parameters

Parameters	Value	Parameters	Value
p	$0, 1/2, 1$	CW	$16, 32, 64$
m	$2, 3, 5$	v_{max}	120 km/h
$E[P]$	8184 bits	MAC	272 bits
PHY	128 bits	ACK	112 bits+ PHY
RTS	160 bits + PHY	σ_{propa}	50 μs
CTS	112 bits + PHY	Lane number	6
SIFS	28 μs	$DIFS$	128 μs

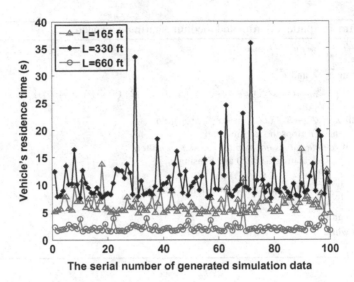

Fig. 3.5 The vehicle's residence time distribution under different coverage ranges

by statistical mean value analysis. By generating 200 times $f_\Gamma^L(\tau)$ value, we can obtain the average residence time considering the six-class traffic flow states and different lengths of coverage range, shown in Fig. 3.6. For example, within $165 * 2$ and $410 * 2$ ft of AP's coverage (i.e., 150 and 250 m), and under service level case E, the mean vehicular residence time is 8.20 s, 13.82 s, respectively. In addition, the vehicular mean residence time is in direct proportion to the traffic flow state. The main reason is that the reduced mean vehicular driving speed prolongs the residence time within a fixed region when the traffic flow state is scaled from the free flow to the congested state. In Fig. 3.7, we can get the mean contending vehicular number under different Traffic flow states within the AP's coverage. For example, under the traffic flow state C, for the length of AP's coverage with 50, 100, 150, 200, and 250 m, the mean contending vehicular number is 4, 8, 11, 14, 17.

3.4.2   Saturated Throughput of IEEE 802.11 Network

The saturated throughput of Drive-Thru Internet is closely related to the probability parameters, such as p_k^{tr} and p_k^{su}, as shown in Figs. 3.8, 3.9, 3.10, and 3.11. Specifically, take IEEE 802.11b and 802.11g for example, Figs. 3.8 and 3.10 show that p_k^{tr} is in direct proportion to the length of AP's coverage that is as the same as we intuitively image. Because with the increase of zone number k, the number of contending vehicles within the zones will be increased, and the probability that at least one transmission will be higher. On the contrary, the probability p_k^{su} that the successful transmission on one considered channel will be therefore reduced

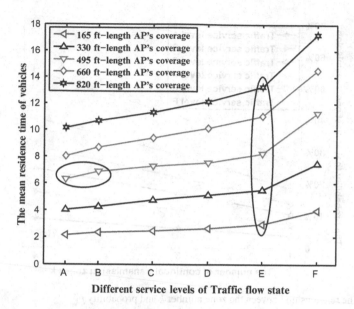

Fig. 3.6 The mean residence time of vehicles under different Traffic flow states

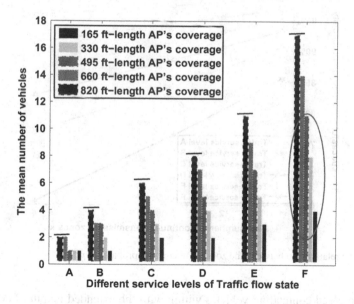

Fig. 3.7 The mean contending vehicular number under different Traffic flow states

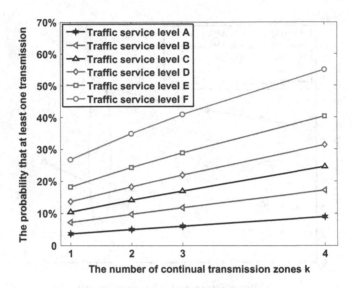

Fig. 3.8 The relationship between the zone number k and probability p_k^{tr}

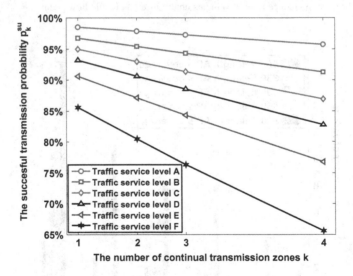

Fig. 3.9 The relationship between the zone number k and probability p_k^{su}

due to increased contending vehicles within with the extended length of coverage. According to the analytical results of IEEE 802.11 networks shown in Figs. 3.12, 3.13, and 3.14, we can see that the optimal mean vehicular throughput can be achieved by selecting best medium access control and sharing region. Specifically, taking the IEEE 802.11a network for example, in Fig. 3.12, the optimal number of medium access control and sharing zones is 5 (i.e., the whole medium access

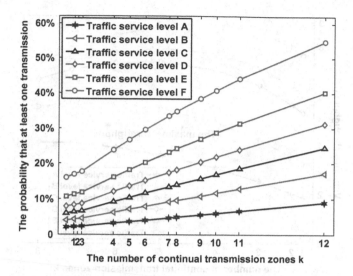

Fig. 3.10 The relationship between the zone number k and probability p_k^{tr}

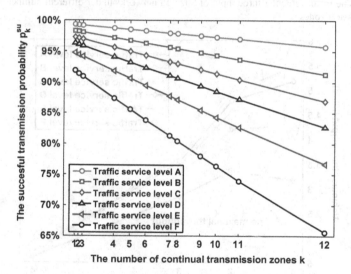

Fig. 3.11 The relationship between the zone number k and probability p_k^{su}

control and sharing range of $2 * \sum_{j=1}^{5} r_{Z_j}$) under the case with traffic service level D. The detailed data for optimal number of medium access control and sharing zones considering different traffic flow states are shown in Table 3.4. Seen from Figs. 3.12, 3.13, and 3.14, an important conclusion can be obtained that we can control the length of medium access and sharing region to cope with the increased contending vehicle number as the traffic becomes congested. On the one hand, with the increased contending vehicular number within medium access control

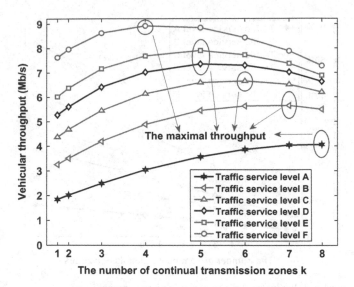

Fig. 3.12 The mean vehicular throughput of 802.11a networks under different number of continual transmission zones

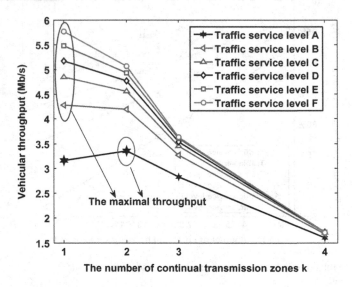

Fig. 3.13 The mean vehicular throughput of 802.11b networks under different number of continual transmission zones

and sharing region, the transmission probability p_k^{tr} on the considered channel is increased, and meanwhile reduce the successful transmission probability. On the other hand, in the multi-rate WLAN, with the extension of medium access and

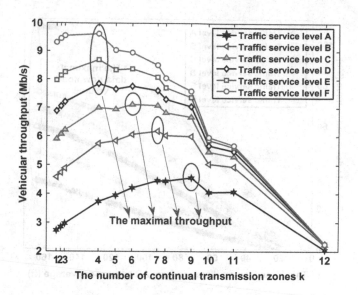

Fig. 3.14 The mean vehicular throughput of 802.11g networks under different number of continual transmission zones

Table 3.4 Optimal medium sharing region for the throughput enhancement of IEEE 802.11 networks

Protocol: optimal range (ft)	Traffic flow state					
	A	B	C	D	E	F
802.11a: range (ft)	330	300	300	260	220	170
802.11b: range (ft)	440	320	320	320	320	320
802.11g: range (ft)	500	210	180	140	140	140

sharing region, the average transmission rate within this coverage will be decreased, which will lead to more time to complete the same size of transmission data volume.

3.4.3 Performance Enhancement

Considering various traffic flow states, we further evaluate the mean transmitted data volume per vehicle within IEEE 802.11 networks. The mean transmitted data volume is not only related to the mean saturated throughput per drive-thru, but also depends on the vehicular sojourn time within AP's coverage. As shown in Figs. 3.15, 3.16, and 3.17, we can see that by optimally choosing the medium access control and sharing region for IEEE 802.11b/g networks, the maximal mean transmitted data volume can be achieved. The detailed performance enhancement of data download volume in IEEE 802.11b/g networks can be shown in Table 3.5. Specially, based on the collected empirical data from the static traffic transport environments in [5], we

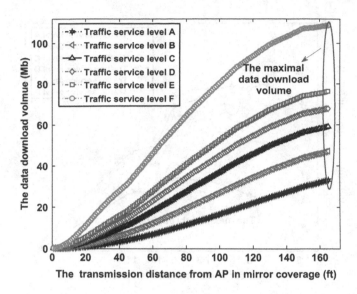

Fig. 3.15 The mean transmitted data volume of 802.11a networks under different access control regions

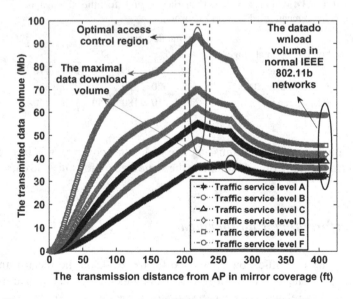

Fig. 3.16 The mean transmitted data volume of 802.11b networks under different access control regions

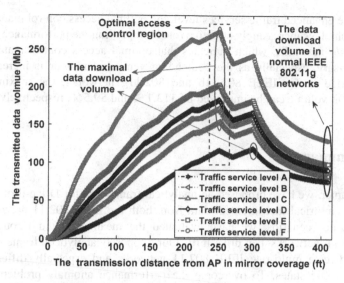

Fig. 3.17 The mean transmitted data volume of 802.11g networks under different access control regions

Table 3.5 Performance comparison (normal vs optimal) for 802.11b/g networks

Protocols: (Mb)	Traffic flow state					
	A	B	C	D	E	F
802.11b (normal)	75.1	82.5	87.1	90.2	97.75	128.9
802.11b (optimal)	118.3	145.3	157.9	183.1	203.3	274.8
802.11g (normal)	32.5	35.9	38.6	41.6	45.6	58.7
802.11g (optimal)	37.7	48.0	55.4	62.1	70.3	93.6

have obtained such simulation results: For traffic flow state A, the optimal medium access range for IEEE 802.11b and IEEE 802.11g networks is $270*2$ ft and $300*2$ ft, respectively. For traffic flow states B, C, D, E, F, the optimal medium access range for IEEE 802.11b and IEEE 802.11g networks is $220*2$ ft and $250*2$ ft, respectively. We summary the conclusions from the simulation results as following:

- The optimal access control management by selecting transmission region within AP's coverage works effectively in IEEE 802.11b/g networks. Due to the short coverage range ($165 * 2$ ft) in IEEE 802.11a networks, it is unnecessary to adjust the transmission region within AP's coverage.
- Observed from the data from Figs. 3.16 and 3.17, we can nearly use a uniform medium access control region to optimize the performance of IEEE 802.11b/g networks for different traffic flow states. Based on the collected empirical data in [5], the uniform medium access control region of IEEE 802.11b/g networks is $220 * 2$ ft and $250 * 2$ ft, respectively.

- With the enhanced traffic services level, the optimal access control management approach plays increasingly effective role in improving the performance of IEEE 802.11b/g networks, which indicates that optimal access control management is necessary, especially as the contending vehicular number is increased. To summarize it, for IEEE 802.11b and 802.11g networks, the maximal data download volume can be improved by 113.1 % and 59.5 %, respectively.

3.5 Summary

In this chapter, we have proposed an analytical framework for Drive-thru Internet. Using the empirical measurement data from both the IEEE 802.11 networks and typical traffic scenarios, we have evaluated the mean saturated throughput of IEEE 802.11 networks per drive-thru. Our proposed analytical framework can adapt to diverse multi-rate IEEE 802.11 networks and especially different real vehicular traffic states. To overcome the performance anomaly problem widely existed in the Drive-thru Internet, we have proposed the optimal access control management approach in the Drive-thru Internet by spatial coordinated medium sharing scheduling. The numerical analysis and performance evaluation showed that the optimal medium access control and sharing region determining can achieve the maximal mean saturated throughput and transmitted data volume. Due to the fully utilized the VANET feature in this proposed approach, i.e., the directed mobility behavior with fair sojourn time within AP's coverage, the optimal access control management approach can simply and efficiently perform both the airtime fairness and drive-thru performance optimization, which can be greatly helpful to boom the Drive-thru Internet.

References

1. Y. Toor, P. Muhlethaler, A. Laouiti, Vehicle ad hoc networks: applications and related technical issues. IEEE Commun. Surv. Tutorials **10**(3), 74–88 (2008)
2. T.H. Luan, X. Ling, X. Shen, Provisioning QoS controlled media access in vehicular to infrastructure communications. Ad Hoc Netw. **10**(2), 231–242 (2012)
3. H. Omar, W. Zhuang, L. Li, Vemac: a tdma-based mac protocol for reliable broadcast in vanets. IEEE Trans. Mob. Comput. **12**(9), 1724–1736 (2013)
4. N. Lu, N. Zhang, N. Cheng, X. Shen, J.W. Mark, F. Bai, Vehicles meet infrastructure: toward capacity-cost tradeoffs for vehicular access networks. IEEE Trans. Intell. Transp. Syst. **14**(3), 1266–1277 (2013)
5. A.D. May, *Traffic Flow Fundamentals* (Prentice Hall, Englewood Cliffs, 1990)
6. D. Zhang, C. Yeo, Enabling efficient wifi-based vehicular content distribution. IEEE Trans. Parallel Distrib. Syst. **24**(3), 233–247 (2013)
7. W.L. Tan, W.C. Lau, O. Yue, T.H. Hui, Analytical models and performance evaluation of drive-thru internet systems. IEEE J. Sel. Areas Commun. **29**(1), 207–222 (2011)

8. J. Harri, F. Filali, C. Bonnet, Mobility models for vehicular ad hoc networks: a survey and taxonomy. IEEE Commun. Surv. Tutorials **11**(4), 19–41 (2009)
9. K. Abboud, W. Zhuang, Modeling and analysis for emergency messaging delay in vehicular ad hoc networks, in *Proceedings of GLOBECOM* (IEEE, Honolulu, 2009), Honolulu, Hawaii, pp. 1–6
10. R. Seide, Capacity, coverage, and deployment considerations for ieee 802.11 g, in *Cisco Systems White Paper*, San Jose, CA, 2005
11. M.J. Khabbaz, W.F. Fawaz, C.M. Assi, A simple free-flow traffic model for vehicular intermittently connected networks. IEEE Trans. Intell. Transp. Syst. **13**(3), 1312–1326 (2012)
12. G. Bianchi, Performance analysis of the ieee 802.11 distributed coordination function. IEEE J. Sel. Areas Commun. **18**(3), 535–547 (2000)

Chapter 4
Cooperative Vehicular Content Distribution in the Drive-thru Internet

Internet and vehicle now representing two most prominent elements of our modern lives, providing in-vehicle Internet access to road travelers has become ever more important [1, 2]. In this chapter, we develop a cooperative Drive-thru Internet framework for large-volume data distribution to vehicles in the highway environment. In specific, note that in the highway scenario, vehicles typically move along a linear topology and drive through the RSU consecutively. We thus propose to form vehicles into a chain cluster so that the cluster members would drive through the RSU consecutively in a sequential order. Each cluster member is responsible to download a non-overlapping part of file, and accordingly with all cluster members moving out the coverage of RSU, an entire file can be downloaded by cluster members but is divided into several parts and separately stored in cluster members. The cluster member then merges the downloaded file to the cluster head. To summarize, ChainCluster virtually extends the connection time of one vehicle to the collectively connection time of a group of vehicles, and therefore enhances the likelihood of intact file download during the short-lived connection time. Moreover, ChainCluster explicitly separates the download phase from RSU from the file merge and recovery phase when the entire cluster leave the RSU coverage. This can fully utilize the precious connection time of vehicles to RSUs. Within the ChainCluster framework, we develop an accurate and simple analytical framework to evaluate the data volume that could be downloaded for each cluster per drive-thru. In particular, our model investigates on the instantaneous download performance of vehicles by deploying a microscopic vehicular mobility model as introduced in [3]. Based on the fundamental IEEE 802.11b MAC scheme, we derive the expression of the overall file download time.

© The Author(s) 2015   45
H. Zhou et al., *Cooperative Vehicular Communications in the Drive-thru Internet*,
SpringerBriefs in Electrical and Computer Engineering,
DOI 10.1007/978-3-319-20454-3_4

The remainder of this chapter is organized as follows. Section 4.1 describes the system model. Section 4.2 presents the cooperative content distribution protocol design. Section 4.3 presents the cooperative content distribution protocol evaluation. Section 4.4 introduces the related deployment insight. Section 4.5 closes the chapter with conclusions.

4.1 System Model

We consider a highway Drive-thru Internet as shown in Fig. 4.1. An RSU is installed on the roadside which serves as the gateway to provide Internet services to vehicles driving through its coverage. Throughout the work, we focus on a tagged vehicle, which subscribes to download a file from the RSU. The file size is considered to exceed the data volume that can be retrieved by the tagged vehicle itself per single drive-thru; otherwise, the problem is trivial. As such, a cooperation scheme is necessary. Further, we define the low-speed driving lane as the cooperative download lane to avoid blocking up the free driving of followed vehicles.[1] With cooperation, the tagged vehicle and other cooperative vehicles can jointly conduct content download and distribution.

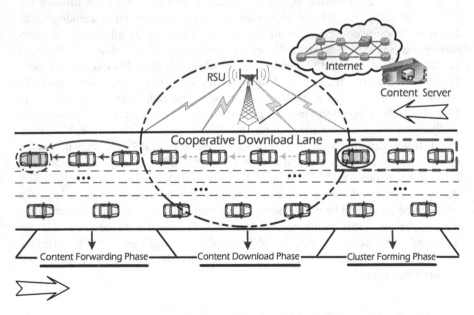

Fig. 4.1 A cooperative content distribution scenario for vehicles in highway Drive-thru Internet

---

[1]Note that as cluster members may be in different lanes, all of them can drive onto the download lane to form a linear chain cluster.

To help the tagged vehicle finish downloading the desired data volume, we form a vehicle cluster to cooperatively download the subscribed file for the tagged vehicle. The cluster appears a linear topology on the highways and each cluster member is responsible to download and submit one separate part of the file to the tagged vehicle. With vehicles consecutively driving through the RSU, the cooperation scheme virtually extends the effective download time of the tagged vehicle to that of a long cluster. By adjusting the number of cluster members, the data volume that can be downloaded by the cluster for each drive-thru is thus controllable.

As shown in Fig. 4.1, the proposed cooperative scheme consists of three phases: cluster forming phase, content download phase, and content forwarding phase. Specifically, if the tagged vehicle would like to download a large-sized file, the tagged vehicle will invite other vehicles followed itself to form a cluster. Once the tagged vehicle enters into the RSU's coverage, the tagged vehicle can evaluate the current saturated download throughput and select a target file. Several vehicles behind it are selected to form a linear cooperative chain cluster, which is called cluster forming phase. We assume the cluster forming time can be neglected for the reduction of tagged vehicle's download throughput. Within RSU's coverage, each vehicle in the cooperative cluster downloads one non-overlapping part of the file sequentially from the RSU, which is content download phase. Outside the RSU's coverage, the tagged vehicle then collects the parts from the cluster members to recover the file and complete the final download procedure.

Due to the mobility of vehicles, the selection of cooperating vehicles and the performance of ChainCluster highly depend on the mobility features of vehicles and the throughput of their connections to the RSU and each other. In what follows, we introduce the vehicle mobility model and the throughput analysis of V2R and V2V communications.

4.1.1 Microscopic Vehicular Mobility

We apply the kinematic equation with some real-world driving constraints to model the microscopic mobility behavior for the free driving of vehicles on highways, shown in (4.1). The introduced microscopic vehicular mobility can be treated as a car-following model with corresponding speed and distance constraints, which are much closer to the real highways vehicular environment and widely used in the literatures [3, 4]. Specially, at the start of each time duration Δt, the vehicle updates its speed based on (4.1), then this vehicle keeps this updated speed during the whole period of Δt.

$$v_i(t + \Delta t) = v_i(t) + \eta_i(t) \cdot a \cdot \Delta t \tag{4.1}$$

where $v_i(t)$ denotes vehicle i's velocity at time t, $\eta_i(t)$ denotes a random adjusting parameter for acceleration or deceleration at time t, a is a constant, and Δt is the updated time unit.

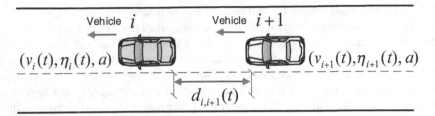

Fig. 4.2 Mobility model of vehicles on the highway

In addition, the vehicle mobility confines to two rules, namely speed constraint and distance constraint, as illustrated in Fig. 4.2.

- *Rule 1 (Speed Constraint)*: The velocity of each vehicle is restricted to the interval $[v_{min}, v_{max}]$ which represents the speed limitation on the road. In this manner, we track the vehicular velocity, denoted by v_i, every Δt period. If $v_i(t + \Delta t) < v_{min}$, then $v_i(t + \Delta t) = v_{min}$; if $v_i(t + \Delta t) > v_{max}$, then $v_i(t + \Delta t) = v_{max}$. In this work, we set $v_{min} = 60$ km/h and $v_{max} = 120$ km/h.
- *Rule 2 (Distance Constraint)*: The distance constraint between any two consecutive vehicles in the same lane includes: minimum inter-vehicle driving distance and maximal inter-vehicle driving distance. The former is referred to as safety distance, denoted by Ψ_{safe}, which depends on the vehicle's realtime velocity. The latter is an appropriate car following distance, denoted by Ψ_{follow}, which can save more space for other driving vehicles. Let $d_{i,i+1}(t)$ denote the distance between two consecutive vehicle V_i and V_{i+1}. The constraint condition is regulated as: if $d_{i,i+1}(t) < \Psi_{safe}$, then $v_{i+1}(t) = v_i(t) - \theta \cdot a$; If $d_{i,i+1}(t) > \Psi_{follow}$, then $v_{i+1}(t) = v_i(t) + \theta \cdot a$. Typically, $0 < \theta \le 1$, Ψ_{safe} is a constant that can be set between 50 and 120 m, and Ψ_{follow} is a constant that can be set to be less than 300 m.

We denote the initial inter-vehicle distances for all vehicles in the cluster by $d_{i,i+1}(t_0) = \Psi_{safe}(1 + \varepsilon_i)$, where ε_i is a random variable uniformly distributed in [0,1]. In the Drive-thru Internet scenario, the vehicle mobility intensely affects the sojourn time within RSU's coverage and data download volume. Therefore, we first investigate the tagged vehicle's sojourn time, and let Γ_i denote the vehicle V_i's sojourn time. Due to the driving speed and inter-vehicle distance restriction, the velocity of vehicles on the road are mutually dependent. For simplification, the tagged vehicle can be regarded as the head in a traffic queue. We provide extensive simulation results in Fig. 4.3–4.5 to numerically show the tracking results of vehicular speed, driving distance and inter-vehicle distance with the proposed mobility model. In Fig. 4.5, we restrict the maximal vehicular inter-distance to be under 260 m. Those results can be used to track the mobility states of individual vehicle in real time, which can be useful in the downlink analysis and Drive-thru Internet performance evaluation. According to (4.1) and the mobility rules, Γ_i is defined as:

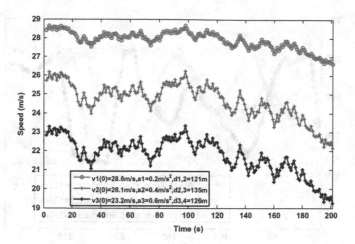

Fig. 4.3 The driving distance tracking

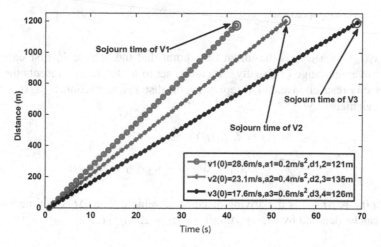

Fig. 4.4 The microscopic mobility tracking under different kinematic equation parameters

$$\Gamma_i \triangleq \left\{ \tau_i \mid \sum_{k=1}^{\lceil \tau_i / \Delta t \rceil} v_i(t_0 + \Delta t \cdot k) \cdot \Delta t = \zeta \right\} \tag{4.2}$$

$$s.t. \qquad v_{\min} \leq v_i(t_0 + \Delta t \cdot k) \leq v_{\max} \tag{4.3}$$

$$\Psi_{safe} \leq d_{i,i+1}(t_0 + \Delta t \cdot k) \leq \Psi_{follow} \tag{4.4}$$

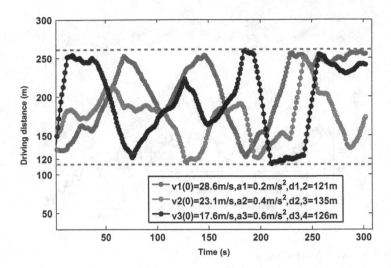

Fig. 4.5 The driving speed tracking

where $v_i(t_0) = v_0$, t_0 is the initial time point that the vehicle V_i first enters the
RSU's coverage range ζ; Usually, Δt can be set to as 1; (4.3) represents the max-
min velocity restriction and (4.4) represents the distance restriction; $d_{i,i+1}(t_0 + k \cdot \Delta t)$
can be calculated by

$$d_{i,i+1}(t_0 + \Delta t \cdot k) = d_{i,i+1}(t_0)$$
$$+ \sum_{q=1}^{k} (s_i(t_0 + \Delta t \cdot q) - s_{i+1}(t_0 + \Delta t \cdot q)) \tag{4.5}$$

where $s_i(t_0 + \Delta t \cdot q)$ is the driving distance of vehicle V_i at $\Delta t \cdot q$th time interval,
which can be denoted by $s_i(t_0 + \Delta t \cdot q) = v_i(t_0 + \Delta t \cdot q) \cdot (\Delta t)$.

4.1.2   V2R Communication

We consider an adaptive V2R transmission rates model [5, 6], in which the RSU's
transmission rates depend on the distance from the RSU to the receiving vehicle. As
shown in Fig. 4.6, the transmission rates received in those divided ranges are denoted
by the rate set $C = \{c_1, c_2, \ldots, c_k, \ldots, c_M\}$, where c_k is the received data rate within
kth zone of RSU's coverage. The adaptive transmission rates within RSU's coverage
are symmetrical, and $M = 7$ in IEEE 802.11b standard. The transmission bit rates
in different zones can be expressed as (4.6). For the ease of illustration, we consider
an ideal MAC protocol similar to [6] in which the RSU airtime is equally allocated
to vehicles in the coverage. Our work can also be easily extended to consider the
more complicated MAC.

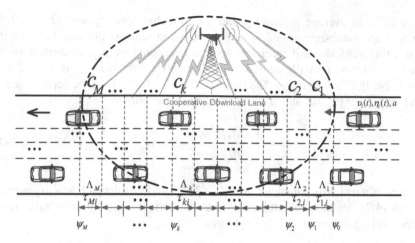

Fig. 4.6 The zone division and adaptive transmission rates

$$c_k(k = 1, \ldots, M) \triangleq \begin{cases} 1\,\text{Mbps}, & k = 1,7 \\ 2\,\text{Mbps}, & k = 2,6 \\ 5.5\,\text{Mbps}, & k = 3,5 \\ 11\,\text{Mbps}, & k = 4 \end{cases} \tag{4.6}$$

4.1.3 V2V Communication

It is necessary to have full knowledge of statistical properties of the V2V channel to analyze the communication performance [7–9]. Here, we adopt the real-world measurement results in [9] to analyze the physical layer capacity of V2V communications. For the typical short-distance connections feature, the distribution of received signal's amplitude in a vehicular receiver gradually transmits from Rician distribution to Rayleigh distribution as the inter-vehicle distance is increased. We model the fast fading highways vehicular environment using the Nakagami-m distribution[2] [9], where the p.d.f. of signal amplitude can be represented by the Nakagami (μ, Ω) distribution as

$$\Im(x, \mu, \Omega) = \frac{2\mu^{\mu}}{\Gamma(\mu)\Omega^{\mu}} x^{2\mu-1} \exp(\frac{-\mu}{\Omega} x^2) \tag{4.7}$$

---

[2]This distribution is also widely used in V2V communication analysis and evaluation, e.g., [10].

where Ω is an average received power in the fading envelop and $\Omega = E(x^2)$, μ is a shape parameter related to the environment and the distance between the transmitting vehicle i and receiving vehicle j, denoted by $d_{i,j}$. According to the measurement result in [9] and the mobility model, if $90.5 \leq d_{i,j} \leq 230.7$, $\mu = 0.74$; if $230.7 < d_{i,j} \leq 588.0$, $\mu = 0.84$, and $\Gamma(\mu)$ can be represented as $\Gamma(\mu) = \int_0^\infty e^{-x} x^{\mu-1} dx$, and Ω can be calculated as follows:

$$\Omega = \frac{P_t G_t G_r h_t^2 h_r^2}{d_{i,j}^\theta L} = \varpi \cdot d_{i,j}^{-\theta} \tag{4.8}$$

where ϖ is a path loss parameter, $\varpi = \frac{P_t G_t G_r h_t^2 h_r^2}{L}$, and θ is the path loss exponent.

With (4.7), we have the probability that the current signal to noise ratio at the receiver is larger than a fixed threshold as

$$\Pr\{\frac{\Omega}{\lambda} > \wp\} = \frac{\Gamma(\mu, \frac{\mu}{\Omega}\lambda\wp)}{\Gamma(\mu)} \tag{4.9}$$

where λ denotes the thermal noise power at the receivers, $\wp$ is a constant threshold, $\Gamma(\mu, \frac{\mu}{\Omega}\lambda\wp)$ is the upper incomplete gamma function, denoted by $\Gamma(\mu, \frac{\mu}{\Omega}\lambda\wp) = \int_{\frac{\mu}{\Omega}\lambda\wp}^\infty t^{\mu-1} e^{-t} dt$.

We assume the wireless transceivers in vehicles can be adapted to support up to K discrete modulation rates based on the current V2V link SNR, denoted by $H = \{\pi_1, \pi_2, \ldots, \pi_K\}$ with $\pi_1 < \pi_2 \ldots < \pi_K$. Specifically, if the current SNR is above the threshold $\wp_k$ and smaller than $\wp_{k+1}$, the modulation rate is set to as π_k, where we set $\wp_{K+1}$ as ∞. As such, based on (4.9), we have the expression that the transmission rate π_k for the current V2V link is selected with the probability,

$$\Pr\{\pi_k\} = \Pr\{\wp_k < \frac{\Omega}{\lambda} < \wp_{k+1}\}$$

$$= \begin{cases} \dfrac{\Gamma(\mu, \frac{\mu\lambda\wp_k}{\Omega}) - \Gamma(\mu, \frac{\mu\lambda\wp_{k+1}}{\Omega})}{\Gamma(\mu)}, & k = 1, .., K-1 \\[3mm] \dfrac{\Gamma(\mu, \frac{\mu}{\Omega}\lambda\wp_k)}{\Gamma(\mu)}, & k = K \end{cases} \tag{4.10}$$

4.2 Cooperative Content Distribution Protocol Design

Due to the limited sojourn time and large number of contending vehicles within RSU's coverage, the download volume of individual vehicle is very limited such that it is hard to enjoy the resource-consuming Internet services such as video streaming. To remedy that, we propose ChainCluster to improve the Drive-thru Internet access service. ChainCluster is featured by its chain feature of vehicular cooperation behavior on the highways. In specific, our proposal is composed of three phases defined as

- *Cluster Forming Phase*: Once entering the RSU's coverage, the tagged vehicle evaluates the performance of cooperative Drive-thru Internet, and selects a number of vehicles to form a linear chain cluster for downloading sequentially from RSU.
- *Content Download Phase*: Within the RSU's coverage, the tagged vehicle allocates the download task according to the vehicular locations and mobility features. Then, the vehicles in the formed cluster download the corresponding parts of the file.
- *Content Forwarding Phase*: Once departure from the RSU's coverage, the tagged node collects the downloads from cluster members to recover the target file.

4.2.1   Cluster Forming Phase

Before entering into the RSU's coverage, the tagged vehicle coordinates to form a chain cluster. Several typical clustering approaches have been investigated in VANET [11–13], where the basic idea about how to find the partners is similar. We focus on how to form a chain topology before arriving at RSU's coverage to keep it stable. To this goal, the tagged vehicle first broadcasts a cooperation request to the followed vehicles in its communication range. Vehicles who are willing to cooperate will make a confirmation to the tagged vehicle. After received the cooperation confirmations from the followed vehicles, the tagged vehicle will ask the confirmed partners to follow one by one after it and drive toward onto the cooperative download lane. Finally, the tagged vehicle needs to collect some simple mobility information of the members in the chain cluster (e.g., initial speed, inter-vehicle distance and position, etc.) to calculate the cluster size. Here, we implicitly assume that all vehicles are equipped with GPS devises to acquire the timely updates of the location information. This is a working assumption as GPS has already been a widely adopted technology, and it is also a key enabling component of vehicular communications.

Once entering the RSU's coverage, the tagged vehicle evaluates the performance of cooperative Drive-thru Internet. The number of vehicles that needs to be requested to join ChainCluster depends on the target file size. If the needed vehicle number is larger than the vehicle number in the formed cluster by using the one-hop broadcasting cooperation messages, the tagged vehicle should adapt the download task to available cluster size. In addition, as the data download volume of a ChainCluster is closely related to the vehicle mobility, the cluster formation should also adapt the time-varying velocity of vehicles. In what follows, we provide the analysis to guide the tagged vehicles to evaluate the throughput of cooperative Drive-thru Internet and the number of vehicles in ChainCluster.

4.2.1.1 Analysis of Contending Vehicle Number

Vehicles within RSU's coverage contend with each other to share RSU's bandwidth. For a vehicle V_i, contending vehicle number (CVN) is defined as the number of vehicles that are concurrently within the RSU's coverage. In order to derive the achieved data download volume by individual vehicle, it is crucial to analyze the instant value of CVN. We derive the CVN of the tagged vehicle V_i at different time points. Note that during the sojourn time within RSU's coverage, V_i's CVN changes with time due to the dynamic arrivals and departures of vehicles to RSU.

We first consider the single-lane highways scenario, and divide the whole coverage range into M zones that is consistent with zones for adaptive data-rate transmission, denoted by $\Lambda = \{\Lambda_1, \Lambda_2, \ldots, \Lambda_M\}$, and shown in Fig. 4.6. Starting from the entrance point marked as ψ_0, the $M + 1$ marked points are denoted by $\psi_0, \psi_1, \ldots, \psi_M$, respectively, which satisfy the condition that $\sum_{k=1}^{M} |\psi_k - \psi_{k-1}| = \zeta$, where ζ is the coverage range of RSU. Further, we set the time when the tagged vehicle arrives at the entrance point ψ_0 as the initial time t_0, and set $t_0 = 0$ in our analysis. At this initial time t_0, there are μ_i vehicles already in the RSU's coverage and κ_i vehicles that followed behind. Let N_{de} and N_{ar} denote the set composed of the μ_i vehicles and the κ_i vehicles, respectively. To calculate the variations in the number of vehicles in RSU's coverage during V_i's sojourn time, we introduce more notations related to the sets N_{de} and N_{ar} as follows.

- N_{de}: Let $\psi_{t_0}^{v_j}$ be the location of the vehicle V_j in the set N_{de} at time t_0, i.e., the location of vehicle V_j as the tagged vehicle V_i just arrived at the entrance point ψ_0. Let $\zeta_{M,t_0}^{v_j}$ be the distance between the location of the vehicle V_j at time t_0 and the departure point ψ_M, denoted by $\zeta_{M,t_0}^{v_j} = |\psi_M - \psi_{t_0}^{v_j}|$. Let $\tau_{V_j}^{de}$ be the time point that the vehicle V_j arrives at the departure point ψ_M, which can be calculated by (4.11). We define $T^{de} = \{\tau_{V_1}^{de}, \tau_{V_2}^{de}, \ldots, \tau_{V_{|\mu_i|}}^{de}\}$, $\mu_i = |N_{de}|$ and $0 \leq \mu_i \leq \lceil \zeta / \Psi \rceil$. We assume the value of μ_i can be obtained via a deployed camera within RSU's coverage.

- N_{ar}: Let $\psi_{t_0}^{v_k}$ be the location of the vehicle V_k in the set N_{ar} at time t_0, and let $\zeta_{0,t_0}^{v_k}$ be the distance between the location of the vehicle V_k at time t_0 and the entrance point ψ_0, denoted by $\zeta_{0,t_0}^{v_k} = |\psi_0 - \psi_{t_0}^{v_k}|$. Let $\tau_{V_k}^{ar}$ be the time point that the vehicle V_k reaches to entrance point ψ_0, which can be calculated by (4.12). The set of driving time for all vehicles in N_{ar} is denoted by $T^{ar} = \{\tau_{V_1}^{ar}, \tau_{V_2}^{ar}, \ldots, \tau_{V_{|\kappa_i|}}^{ar}\}$, $\kappa_i = |N_{ar}|$, and κ_i is the number of possible cooperative vehicles after the tagged vehicle's request, which depends on the factual scenario.

$$\tau_{V_j}^{de} \triangleq \{\tau_j \mid \sum_{q=1}^{\lceil \tau_j / \Delta t \rceil} v_j(t_0 + \Delta t \cdot q) \cdot \Delta t = \zeta_{M,t_0}^{v_j}\} \tag{4.11}$$

$$\tau_{V_k}^{ar} \triangleq \{\tau_k \mid \sum_{q=1}^{\lceil \tau_k / \Delta t \rceil} v_k(t_0 + \Delta t \cdot q) \cdot \Delta t = \zeta_{0,t_0}^{v_k}\} \tag{4.12}$$

where ψ_M and ψ_0 are the departure point and the entrance point of RSU's coverage, respectively.

For the tagged vehicle V_i, we can get the set of individual running time point T^{tag} in the M zones of RSU's coverage, denoted by $T^{tag} = \{\tau_{1,i}, \tau_{2,i}, \ldots, \tau_{M,i}\}$, where $\tau_{k,i}$

denotes the time point that the vehicle V_i leaves the kth zone , and can be calculated by (4.13).

$$\tau_{k,i} \triangleq \{\tau_k \mid \sum_{q=1}^{\lceil \tau_k/\Delta t \rceil} v_i(t_{k-1,i} + \Delta t \cdot q) \cdot \Delta t = \zeta_{k,0}^{v_i}\} \qquad (4.13)$$

The data download volume by the tagged vehicle V_i per drive-thru is closely related to the variations of CVN in M zones of RSU's coverage, which only depends on the number of vehicles at every arrival time point and departure time point. To analyze the effects, we process the values calculated by (4.11)–(4.13) by three steps, introduced as follows:

Step 1: The three time sets are combined into a large set T, i.e., $T = \{T^{de}, T^{ar}, T^{tag}\}$.

Step 2: We truncate the set T to make its elements are within the fixed interval $\tau_{0,i}$, $\tau_{M,i}$, where $\tau_{0,i}$ and $\tau_{M,i}$ are the time point that the tagged vehicle V_i enters into and leaves from the RSU's coverage, respectively. For simplicity, we let $\tau_{0,i} = 0$.

Step 3: We chronologically reorder all elements in set T and denote the reordered set as $T'=\{\tau_1, \tau_2, \ldots, \tau_W\}$, where W is the number of elements in set T'.

To calculate the variations of CVN during the tagged vehicle's sojourn time, we design a moving time window to analyze the changing states in the set T', shown in Fig. 4.7, and the size of time window is defined as $\Im_k = |\tau_k - \tau_{k-1}|, k = 1, 2, \ldots, W$. In fact, the sizes of W time windows are not the same, which can reflect each changing state for the contending number in M zones. To judge whether there exists a variation of CVN for the tagged vehicle V_i during the period of kth time window $\Im_k$, we define a decision function, which reflects whether one vehicle belonging to the set $N_{de} \cup N_{ar}$ leave or arrive at the RSU's coverage. If one vehicle arrives at the RSU's coverage, the current CVN will be added by one, and vice versa.

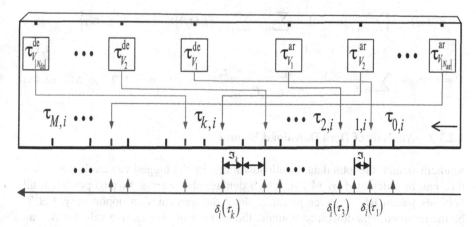

Fig. 4.7 The illustration of time sequence distribution for a tagged vehicle and the vehicles driving in the RSU's coverage

The decision function can decide the changing state of CVN at each discrete time point τ_k for the tagged vehicle V_i, which is denoted by $\delta_i(\tau_k)$, where $k = 1, 2, \ldots, W$.

$$\delta_i(\tau_k) = \begin{cases} -1, & \tau_k \in \mathrm{T}^{de} \\ 0, & else \\ +1, & \tau_k \in \mathrm{T}^{ar} \end{cases} \tag{4.14}$$

According to (4.14), during the tagged vehicle V_i's sojourn time on single-lane highways, we can easily formulate the variations of CVN at individual time point τ_k in the set of T', denoted by $N_i(\tau_k)$. Further, we can calculate the value of $N_i(\tau_k)$ in M different zones, i.e., at discrete time point τ_k, the CVN for the tagged vehicle V_i driving within nth zone can be calculated, which is denoted by $N_i^n(\tau_k)$, $k = 1, 2, \ldots, W$ and $n = 1, 2, \ldots, M$.

$$N_i(\tau_k) = \mu_i + 1 + \sum\nolimits_{j=1}^{k} \delta_i(\tau_j), \tau_k \in \mathrm{T}' \tag{4.15}$$

$$N_i^n(\tau_k) = \{N_i(\tau_k) \,|\, \tau_{n-1,i} \leq \tau_k \leq \tau_{n,i}\} \tag{4.16}$$

Our analysis of CVN for a single-lane highways scenario can be easily extended to the multi-lane highway scenario. For a multi-lane scenario with the highway lanes of P ($P = 2, 4, 6,$ or 8), the extended decision function to judge the variations in CVN for vehicle V_i driving in nth zone can be derived by (4.17). However, in real applications, the traffic flows in different lanes are independent. For a tagged vehicle, it will be hard to collect enough information of traffic flow in different lanes for the analysis of CVN. For this consideration, we estimate the total CVN in multi-lane scenario simply based on the result of single-lane. For instance, we simply use $N_i^n(\tau_k) \cdot P$ to estimate the variations in the CVN for the tagged vehicle V_i at the time τ_k in multi-lane scenario, where P is the number of highway lane.

$$N_i^n(\tau_k) = \left\{\sum\nolimits_{j=1}^{P} \mu_{i,j} + 1 + \sum\nolimits_{j=1}^{P} \sum\nolimits_{r=1}^{k} \delta_{i,j}(\tau_r) \,\Big|\, \tau_{n-1,i} \leq \tau_k \leq \tau_{n,i}\right\},$$
$$\tau_k \in \mathrm{T}', n = 1, \ldots, M. \tag{4.17}$$

$$\Phi_{v_i} = \sum\nolimits_{k=1}^{W} \frac{\Im_k \cdot \{c_n \,|\, \tau_{n-1,i} \leq \Im_k \leq \tau_{n,i}\}}{N_i^n(\tau_k) \cdot P}, v_i \in \Re, n = 1, 2, \ldots, M \tag{4.18}$$

4.2.1.2 Analysis of Data Download Volume

Mathematically, the total data download volume by the tagged vehicle V_i per drive-thru can be calculated by (4.18), which denotes the average throughput when all vehicles within RSU's coverage fairly share the transmission opportunity. Let Υ be the required data download volume, the number of cooperative vehicles N_c can be calculated by (4.19). In a practical case, we can quickly estimate the number of required cooperative vehicles simply by $N_c \approx \lceil \Upsilon / \Phi_{v_i} \rceil$.

The expression of Eq. (4.19) is shown as

$$N_c = \left\{ \min\{\gamma\} \,|\, \gamma = 1, 2, \ldots, \kappa_i, \sum_{i=1}^{\gamma} \Phi_{v_i} \geq \Upsilon \right\} \tag{4.19}$$

4.2.2 Content Download Phase

According to the analysis in the cluster forming phase, the tagged vehicle will select $N_c - 1$ vehicles that follow behind it and have similar mobility for cooperative content download. Once the tagged vehicle drives into RSU's coverage, it will subscribe to a file. To enable multi-party download, the file is divided into N_c chunks based on the evaluated download throughput of individual vehicle. The ongoing download content chunks are marked by $\{1, 2, \ldots, N_c\}$, which is corresponding to the download file for the vehicles $V_1, V_2, \ldots, V_{N_c}$, respectively. Figure 4.8 shows an example in which the chain cluster is composed of four vehicles, including the tagged vehicle at the cluster head. Once received the application acknowledgement from RSU, all the vehicles driving into RSU's coverage can download the corresponding allocated tasks.

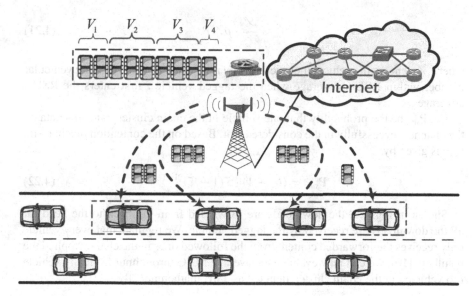

Fig. 4.8 The content division and cooperative download illustration

4.2.3　Content Forwarding Phase

The ChainCluster adopts the contention-based MAC for distributed downloads forwarding on highways. We first evaluate the throughput of chain cluster under cooperative content forwarding strategy. We consider that the vehicles apply the IEEE 802.11b DCF (distributed coordination function) scheme for MAC scheduling in ChainCluster. In addition, each packet is transmitted by means of RTS/CTS mechanism to eliminate the hidden terminals problem. We consider a constant contention window size for counting the backoff time, denoted as CW. Let $\mathfrak{T}$ denote the average transmission probability of each vehicle, and

$$\mathfrak{T} = \frac{2}{CW + 1}. \tag{4.20}$$

Different from that within RSU's coverage, the analysis of transmissions contention outside RSU depends on the vehicles' carrier sensing range. To calculate the number of vehicles which are contending the transmission channel in the vehicle transmission range, according to the vehicle number within RSU's coverage and the proportion to vehicles' carrier sensing range, we can estimate the vehicular number in the tagged vehicles' carrier sensing range, $\mathfrak{N}$ can be given as

$$\mathfrak{N} \approx \frac{\mathscr{S}}{\zeta} \cdot \mu_i \cdot P \tag{4.21}$$

where $\mathscr{S}$ denotes the vehicular carrier sensing range, and μ_i is the existing vehicular number within RSU's coverage when the tagged vehicle i first enters the RSU's coverage.

Let $\mathbf{P}_{suc}$ be the probability that one vehicle in the chain cluster transmits data on the channel successfully in the considered slot. Based on the contention mechanism, $\mathbf{P}_{suc}$ is given by

$$\mathbf{P}_{suc} = (N_c - 1) \cdot \mathfrak{T}(1 - \mathfrak{T})^{\mathfrak{N}-1} \tag{4.22}$$

Shown in Fig. 4.9, the downloads are forwarded from the back to the head till all the downloads are received by the tagged vehicle. We regulate that every vehicle only receives the forwarded content from the followed ones in the chain. Applied the results in [10], we can now evaluate the average MAC throughput from any vehicle i to vehicle j in the chain cluster, denoted as $\mathbf{Thr}$, mathematically,

$$\mathbf{Thr} = \frac{E[X_{payload}]}{E[Y_{slot-time}]}$$
$$= \frac{P_{suc}*PA}{\underbrace{P_{em}*\sigma}_{E[T_{empty}]} + \underbrace{P_{su}*T_{su}}_{E[T_{tran}]} + \underbrace{(P_{tr} - P_{su})*T_{co}}_{E[T_{coll}]}} \tag{4.23}$$

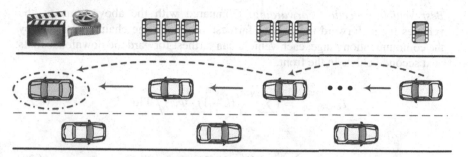

Fig. 4.9 The downloads forwarding in a vehicular chain-cluster

where $E[X_{payload}]$ denotes the average payload volume in the chain cluster transmitted successfully in a slot time, PA is the package sizes including the package head. $E[Y_{slot-time}]$ denotes the average time slot length in DCF scheme, including the average empty slot time $E[T_{empty}]$, the average successful transmission time $E[T_{tran}]$, and the collision time $E[T_{coll}]$. Based on the DCF performance analysis in [14], they can be given by (4.24), (4.25), (4.26), respectively,

$$\mathbf{P}_{em}\sigma = (1 - \mathfrak{T})^{\mathfrak{N}}\sigma_{SlotTime} \tag{4.24}$$

$$\mathbf{P}_{su}T_{su} = \mathfrak{N}\mathfrak{T}(1 - \mathfrak{T})^{\mathfrak{N}-1}(RTS + 3SIFS + 4\sigma_{propa}$$
$$+ CTS + ACK + DIFS + \frac{E(F)}{E(C)}) \tag{4.25}$$

$$(\mathbf{P}_{tr} - \mathbf{P}_{su})T_{co} = ((1 - (1 - \mathfrak{T})^{\mathfrak{N}}) - \mathfrak{N}\mathfrak{T}(1 - \mathfrak{T})^{\mathfrak{N}-1})(RTS + DIFS + \sigma_{propa}) \tag{4.26}$$

In (4.25), $E(F)$ is the average frame length and $E(C)$ is the average modulation rate. Considering the maximal communication range of vehicles (300 m) and the inter-vehicle distance restriction, the vehicle can only connect with maximal two vehicles in the communication range. To calculate the downloads forwarding delay for the whole chain cluster, we can consider two strategies to forward the individual downloads, namely, one-by-one forwarding and the best vehicle selection forwarding. In an N_c-vehicle chain cluster, the N_c vehicles are ordered by $1, 2, \ldots, N_c$. According to the hop number of individual vehicle in the formed cluster, we can estimate the downloads forwarding time to retrieve a targeted file, denote as $T_{\text{one-by-one}}$ and T_{best}, respectively. The downloads forwarding strategies in the chain cluster are in the following,

– *One-by-One Forwarding*: For this strategy, the downloads by individual vehicle are forwarded from the rear to the head, and the forwarding route for each vehicle is to forward the downloads to its next first vehicle in the front.

– *Best Vehicle Selection Forwarding*: Compared with the above strategy, the vehicles try to forward the data to farthest vehicle in the chain. Restricted by the communication range, each vehicle can farthest forward the downloads to its next second vehicle in the front.

$$T_{\text{one-by-one}} = \left(\sum\nolimits_{k=1}^{N_c} (k-1) \cdot \Phi_{v_k} \right) \Big/ \mathbf{Thr} \tag{4.27}$$

$$T_{\text{best}} = \left(\sum\nolimits_{k=1}^{N_c} \lceil (k-1)/2 \rceil \cdot \Phi_{v_k} \right) \Big/ \mathbf{Thr} \tag{4.28}$$

4.3 Cooperative Content Distribution Protocol Evaluation

We conduct simulations to evaluate the performance of ChainCluster in highways scenarios using Matlab. The simulation settings, including the vehicular traffic parameters, V2R communication parameters and V2V communication parameters, are specified in Tables 4.1 and 4.2.

Table 4.1 The DSRC data rate and SNR threshold setting

SNR threshold (dB)	5	6	8	11	15	20	25	N/A
Date rate (Mbps)	3	4.5	6	9	12	18	24	27

Table 4.2 Setting of DCF and highways scenario parameters

Parameters	Value	Parameters	Value
P_t	23 dBm	$G_t = G_t$	1
L	1	$h_t = h_t$	1
CW	16, 32, 64, 128	S	500 m
P	2,4,6	λ	−88 dBm
PA_i	1024,2048,5012 bits	H_i	400 bits
σ_{propa}	1 µs	$\sigma_{SlotTime}$	50 µs
$SIFS$	28 µs	$DIFS$	128 µs
ACK	37 µs	CTS	37 µs
v_{min}	60 km/s	v_{max}	120 km/s
η	[−1, 1]	ζ	[600, 1000] m
$d_{i,i+1}^0$	[120, 300]	RTS	53 µs

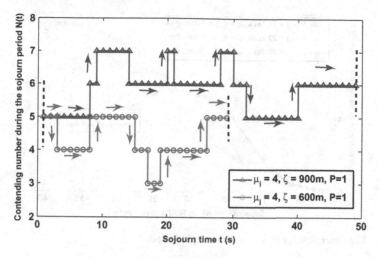

Fig. 4.10 The variations of contending vehicular number

4.3.1 Contending Vehicular Number

Figure 4.10 shows the variations of CVN within RSU's coverage, which depends on current traffic state, including the vehicular driving speed, inter-vehicle distance, acceleration/deacceleration, the number of highways lanes and RSU's coverage. In the simulation, we set a simulation environment with ten vehicles in a line and they will drive through the RSU's coverage, the other driving related parameters are shown as following: the initial driving speed distribution is (rand(1) + 1)*16.67; the inter-vehicle distance distribution is $(rand(1) * 0.25 + 1) * 120$; the acceleration/deceleration distribution is $\pm rand(1)$; the maximal and minimal speed constraints on highways is 16.67 m/s and 33.33 m/s, respectively. As we can see, with the coverage of RSU to be 600 m and 900 m, respectively, the CVN distribution at different sojourn time points are different, attributing to the enlarged RSU coverage. Considering a single-lane scenario, and the existing vehicular number is 4 when a tagged vehicle first drives into the RSU's coverage, the CVN distribution is among [3, 5] and [5, 7], respectively.

4.3.2 Data Download Volume and Cooperative Vehicle Number

Figures 4.11, 4.12, and 4.13 show the impacts of different parameters on the data download volume for a tagged vehicle per drive-thru. From Fig. 4.9, we can observe that with the slowdown of vehicular velocity, the total downloads volume increases. The main reason is that the total sojourn time within RSU's coverage is prolonged. Interestingly, due to the adaptive transmission rates within RSU's

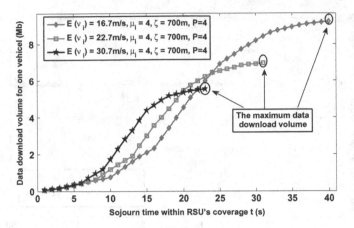

Fig. 4.11 The illustration for downloads under typical speeds

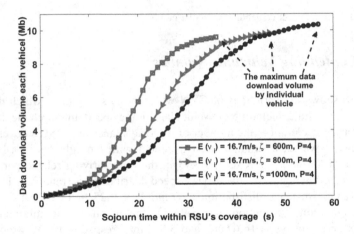

Fig. 4.12 The illustration for data downloading with different coverage

coverage, Fig. 4.11 also shows the varied data download volume in different RSU's zones, e.g., during the middle interval of sojourn time, the vehicle can download more data than the time points in the entrance part and departure part, which can be helpful to make strategies to download the maximal data download volume via controlling the vehicular speed in different zones. Figure 4.12 shows the impact of the size of RSU's coverage on the data download volume. We can see that a larger RSU's coverage doesn't contribute much to the increase of data download volume. When the RSU's coverage is enlarged by 67 % from 600 to 1000 m, the download volume is only increased by nearly 8 %. The main reason is the increased CVN as the coverage range is enlarged. Figure 4.13 shows the impacts of the vehicular velocity and RSU's coverage on the data download volume of individual vehicle in different typical traffic scenarios. When the average driving speed for a tagged

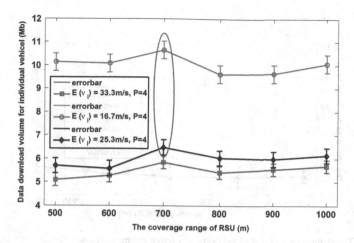

Fig. 4.13 The relationship among data download volume, speed, and coverage

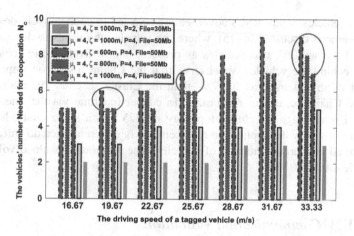

Fig. 4.14 The cooperative vehicular number under typical scenarios

vehicle is 16.7 m/s, 25.3 m/s, and 33.3 m/s, respectively, the maximal data download volume can reach to 10.65 MB, 6.48 MB, 5.84 MB, respectively.

Figure 4.14 shows the cooperative vehicular number under some typical traffic and download scenarios. We observe that with the increase of mobility speed, the more vehicular number will be needed for cooperation, mainly due to the decreased sojourn time within RSU's coverage. For example, to download 50 MB file in Drive-thru Internet, the typical cooperative vehicle number value is among [5, 9] for the four-lane highways. In addition, we also observe that the RSU's coverage makes little contribution to the data download volume at most of time.

Figure 4.15 shows the comparison of the total download volume by five coopera-tive vehicles in the proposed schemes, namely ChainCluster, Cluster, and V2VR. Specially, the Cluster scheme is investigated in [11–13], where several mutual

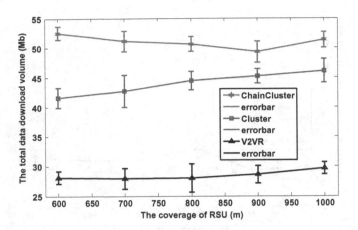

Fig. 4.15 The comparison of download performance of the proposed schemes

connected vehicles within RSUs coverage are selected to cooperatively download. V2VR scheme is discussed in [15], where the vehicle who wants to download the large-sized file will ask two vehicles as proxies: one in front of the vehicle and the other behind the vehicle. We observe that the ChainCluster can improve the download volume as more as 20.87 % than the Cluster scheme. The main reason is that the ChainCluster can maximize the connection time via the linear chain topology. Even though the cluster topology of V2VR is a linear chain, however, V2VR cannot satisfy the application requirements of different sized downloads due to the limited cooperative number of vehicles. The maximal download volume of V2VR is less than 30 MB.

4.3.3 V2V Communication Validation

Figures 4.16 and 4.17 show the process of adaptive data-rate selection and the different data-rate selection probabilities for free driving vehicles, respectively. Figure 4.16 shows that the selected transmission rate varies from 4.5 to 18 Mbps. Figure 4.17 shows the different transmission rates selection probabilities for driving vehicles. The adaptive data-rate selection process and selection probabilities are closely related to the inter-vehicles driving distance on highways, which can also effect the downloads forwarding throughput for ChainCluster.

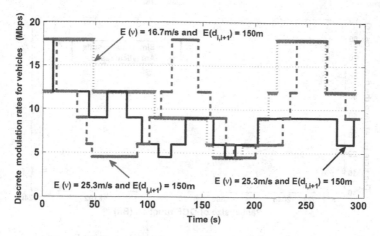

Fig. 4.16 The illustration of adaptive data-rates selection

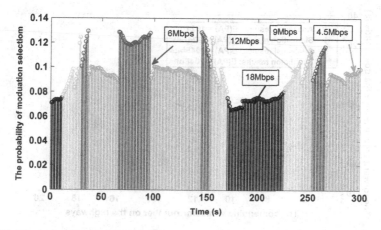

Fig. 4.17 The probability of different data-rates selection

4.3.4 Cooperative Download Forwarding Time

We further demonstrate the impacts of various parameters on the download forward time. We simulate a case that five vehicles form a cluster and cooperatively download the file with 50 MB size. After leaving the RSU's coverage range, the cooperating vehicles will forward their received data (i.e., 10 MB data for each vehicle in this simulation) to the tagged vehicle (i.e., the vehicle at the head of chain cluster). Figure 4.18 shows the impacts of packet size on the download forward time. It is observed that with the increase of effective packet size, the time taken to retrieve a completed file decreases, which is due to the increase of download forwarding throughput. For instance, the packet size increases from 512 bits to 1024 bits, the downloads forwarding time can be decreased by nearly 83.67 %, and the

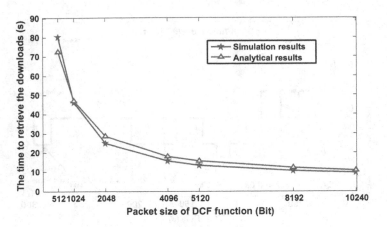

Fig. 4.18 The analytic results and simulation comparison for data forwarding process

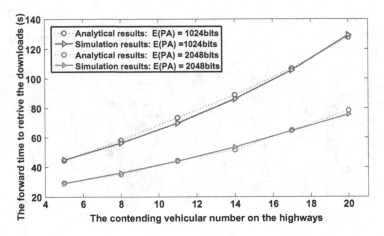

Fig. 4.19 The effect of contending vehicular number for data forwarding delay

simulation results of the whole file retrieving time match with the analytical values very well. Figure 4.19 shows how the CVN affects the downloads forwarding time for retrieving a 50 MB file. We consider a Poisson distribution of vehicular number ranged from 5 to 20 (equally, from the single-lane to four-lane highways scenario). Obviously, the increase of CVN can prolong the downloads forwarding delay. For the case with 1024-bit package size, the downloads forwarding time is increased from nearly 45–129 s. Figure 4.20 shows how the backoff window size in IEEE 802.11 DCF function affects on the downloads forwarding time for retrieving a 50 MB file. Seen in Fig. 4.20, we can optimize the downloads forwarding time by choosing the optimal backoff window size, and in this scenario, the best window size is 32.

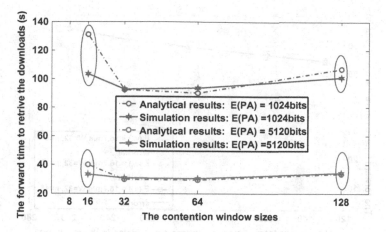

Fig. 4.20 The effect of contending window size for data forwarding delay

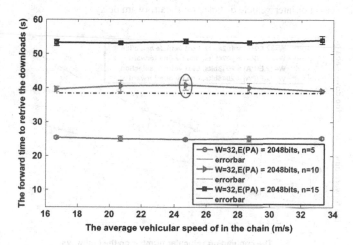

Fig. 4.21 The effect of vehicular driving speed for data forward delay

Figures 4.21, 4.22, and 4.23 demonstrate the impacts of vehicular speed and inter-vehicle distance on the forwarding time. Figure 4.21 shows that the vehicular driving speed affects less on the downloads forwarding time. The main reason is that the increase of driving speed cannot affect the selection of physical transmission data-rate, it makes less contribution to the data forwarding throughout the transmission link. Figure 4.22 shows that with the decrease of inter-vehicle driving distance, the downloads forwarding delay is reduced. When the average inter-vehicle driving distance is enlarged from 130 to 260 m, the downloads forwarding delay is decreased by 11.1 s, 13.4 s, and 11.6 s, respectively. Figure 4.23 shows the comparison of two download forward strategies: "one by one forward" and "best vehicle selection." We observe that the latter can reduce the downloads

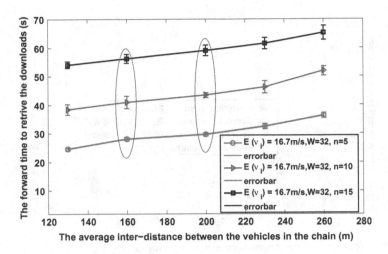

Fig. 4.22 The effect of inter-vehicle distance for data forward delay

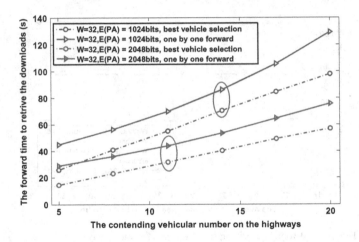

Fig. 4.23 The comparison of two considered forward strategies for data forward delay

forwarding delay. Specially, we set a 300 m communication range of individual vehicle, and for the inter-vehicle distance restriction, each vehicle can connect at most next two vehicles in the linear chain. If the data forwarding route is to find the further inter-vehicle distance each hop, for 1024-bit and 2048-bit package size settings, the downloads forwarding delay can be reduced by nearly 31 and 18 s at most. Figure 4.23 indicates that with a high contention environment (i.e., the large value of CVN), the best vehicle selection will achieve a larger reduction of download forward time.

4.4 Deployment Insight

The ChainCluster scheme provides a fundamental solution to the cooperative content download and distribution for highways vehicular communications, which shows its rationality, necessity, extendibility, and safety and efficiency in reality.

- *Rationality*: We consider that not all vehicles on the highways need to connect to the Internet every time meeting a Drive-thru Internet, so there exists potential vehicles to support cooperation in the ChainCluster scheme.
- *Necessity*: We consider the current data download volume of individual vehicle per drive-thru cannot satisfy the applications for large-sized file download, the ChainCluster scheme can at least support the resource-consuming service of one specific vehicle via utilizing other idle vehicles' download resource.
- *Extendibility*: The ChainCluster scheme can support the P2P sharing scenario for highways VANET, i.e., the tagged vehicle in the chain can be extended to one recommended representative with the same download interest and become an energetic mobile seed to provide the download for neighboring vehicles. The targeted download objective can be one selective popular file.
- *Safety and Efficiency*: The cooperative download lane can avoid blocking up the traffic of followed vehicles for free driving. The cooperative download rules can be implemented on the special lane to improve the throughput in inter-vehicle communications with safer guarantee, e.g., speed control and inter-vehicle distance adjustment, etc.

4.5 Summary

In this chapter, we have proposed and analyzed the ChainCluster scheme for improving the Drive-thru Internet access service. The proposed ChainCluster scheme is a three-phase systematic solution for the cooperative content download and distribution among high-speed vehicles on highways. We have applied a microscopic mobility model, jointly considering two realistic mobility rules to analyze the CVN of a tagged download vehicle, which is crucial to evaluate the cooperative download performance in Drive-thru Internet. We have theoretically derived the data download volume by the tagged vehicle per drive-thru, and the results can be extended to multi-lane highways scenarios. Furthermore, we have proposed a downloads forwarding strategy for the ChainCluster based on the IEEE 802.11 DCF mechanism. From the perspective of real applications, we have derived the downloads forwarding time for different targeted downloads. Via the performance evaluation, we have observed that the vehicular velocity has significant impact on the data download volume whereas the RSU's coverage only has slight effect on it. In addition, to quickly retrieve a file for the subscribed vehicle, we can achieve it by setting an optimized backoff window size, reducing the inter-vehicle distance under the safety distance restriction, and choosing the furthest vehicle in the transmission range for data forwarding.

References

1. L. Yang, F.-Y. Wang, Driving into intelligent spaces with pervasive communications. IEEE Intell. Syst. **22**(1), 12–15 (2007)
2. N. Cheng, N. Lu, N. Zhang, X.S. Shen, J.W. Mark, Vehicular wifi offloading: challenges and solutions. Veh. Commun. **1**(1), 13–21 (2014)
3. F. Bai, N. Sadagopan, A. Helmy, The IMPORTANT framework for analyzing the impact of mobility on performance of RouTing protocols for adhoc NeTworks. Ad Hoc Netw. **1**(4), 383–403 (2003)
4. A.D. May, *Traffic Flow Fundamentals* (Prentice Hall, Englewood Cliffs, 1990)
5. T.H. Luan, X. Ling, X. Shen, Mac in motion: impact of mobility on the mac of drive-thru internet. IEEE Trans. Mob. Comput. **11**(2), 305–319 (2012)
6. W.L. Tan, W.C. Lau, O. Yue, T.H. Hui, Analytical models and performance evaluation of drive-thru internet systems. IEEE J. Sel. Areas Commun. **29**(1), 207–222 (2011)
7. X. Cheng, C.-X. Wang, B. Ai, H. Aggoune, Envelope level crossing rate and average fade duration of nonisotropic vehicle-to-vehicle ricean fading channels. IEEE Trans. Intell. Transp. Syst. **15**(1), 62–72 (2014)
8. X. Cheng, Q. Yao, M. Wen, C. Wang, L. Song, B. Jiao, Wideband channel modeling and intercarrier interference cancellation for vehicle-to-vehicle communication systems. IEEE J. Select. Areas Commun. **31**(9), 434–448 (2013)
9. L. Cheng, B.E. Henty, D.D. Stancil, F. Bai, P. Mudalige, Mobile vehicle-to-vehicle narrow-band channel measurement and characterization of the 5.9 ghz dedicated short range communication (dsrc) frequency band. IEEE J. Select. Areas Commun. **25**(8), 1501–1516 (2007)
10. T.H. Luan, X. Shen, F. Bai, Integrity-oriented content transmission in highway vehicular ad hoc networks, in *Proceedings of IEEE INFOCOM* (2013), pp. 2562–2570
11. H. Su, X. Zhang, Clustering-based multichannel mac protocols for qos provisionings over vehicular ad hoc networks. IEEE Trans. Veh. Technol. **56**(6), 3309–3323 (2007)
12. Z. Wang, L. Liu, M. Zhou, N. Ansari, A position-based clustering technique for ad hoc intervehicle communication. IEEE Trans. Syst. Man Cybern. C Appl. Rev. **38**(2), 201–208 (2008)
13. Y.-C. Lai, P. Lin, W. Liao, C.-M. Chen, A region-based clustering mechanism for channel access in vehicular ad hoc networks. IEEE J. Select. Areas Commun. **29**(1), 83–93 (2011)
14. G. Bianchi, Performance analysis of the ieee 802.11 distributed coordination function. IEEE J. Select. Areas Commun. **18**(3), 535–547 (2000)
15. J. Zhao, T. Arnold, Y. Zhang, G. Cao, Extending drive-thru data access by vehicle-to-vehicle relay, in *Proceedings of ACM VANET* (2008), pp. 66–75

Chapter 5
Conclusion and Future Research Directions

In this chapter, we summarize the main concepts and results presented in this monograph and highlight future research directions. The remainder of this chapter is organized as follows. Section 5.1 presents the concluding remarks. Section 5.2 introduces the potential future works.

5.1 Concluding Remarks

The advance of wireless communications in the recent years have driven the ubiquitous Internet access, especially in-vehicle and on-board vehicular Internet access as people now spend more and more time in cars. This monograph investigated cooperative Drive-thru Internet communication researches can provide guidance for the cost-effective, easy access, and convenient vehicular applications, especially for the delay-tolerant vehicular content distribution.

For the first research issue, we aim at the cooperative vehicular access optimization in the Drive-thru Internet to improve the vehicular access performance. We propose the vehicle-oriented WLAN performance analysis and evaluation model in Drive-thru Internet, and accordingly, the optimal vehicular access approach is proposed. Specially, different from the WLAN performance analysis for static users, we establish a saturated throughput evaluation approach for the widely existing IEEE 802.11 networks (IEEE 802.11a/b/g) based on the empirical traffic flow data, which is more suitable for the real vehicular communication environments. In addition, based on the real Wi-Fi measurement and traffic flow data, we propose an optimal WLAN vehicular access control approach. The proposed access control scheme can be directly applicable to off-the-shelf IEEE 802.11 APs and has the advantage of simplicity and flexibility for the real operation. Moreover, through

© The Author(s) 2015
H. Zhou et al., *Cooperative Vehicular Communications in the Drive-thru Internet*,
SpringerBriefs in Electrical and Computer Engineering,
DOI 10.1007/978-3-319-20454-3_5

both theoretical analysis and extensive simulation results, we show that our proposal can ensure the airtime fairness for medium sharing and boost the throughput performance of Drive-thru Internet.

For the second research issue, we discuss the dynamic cooperative communication in Drive-thru Internet for highways vehicular content distribution. We propose ChainCluster, a cooperative Drive-thru Internet scheme. ChainCluster can select appropriate vehicles and vehicular number based on the content distribution requirement to form a linear cluster on the highway. The cluster members then can cooperatively download the same content file, with each member retrieving one portion of the file, from the roadside infrastructure. With cluster members consecutively driving through the roadside infrastructure, the download of a single vehicle is virtually extended to that of a tandem of vehicles, which accordingly enhances the probability of successful file download significantly. With a delicate linear cluster formation scheme proposed and applied, in this work we first develop an analytical framework to evaluate the data volume that can be downloaded using cooperative drive-thru.

5.2   Potential Future Works

With the predictable booming of Internet-integrated vehicle services, especially the media-rich vehicular Internet access services on the go, an extensive body of researches have been devoted to enabling the cost-effective, high-quality, and flexible vehicular communication networking. IEEE is developing some new WLAN standards, representatively IEEE 802.11af [1, 2] and IEEE 802.11ac standard [3, 4], which can support either high throughput or longer-range connected vehicular Internet access services on the go [5, 6]. Specifically, IEEE 802.11af standard operates on the TV white spectrum, which owns more than three times spectrum resource of ISM band and can transmit at least three or four times coverage range of a common IEEE 802.11 wireless local area network, and generally with an attractive name as "Super WiFi" or "WhiteFi." IEEE 802.11ac is the fifth generation WiFi networking standard and can provide super speed of WiFi links exceeding 1 GB/s, which can meet the needs of bandwidth-hungry wireless applications such as large file transfer, high definition video streaming, wireless display, and cellular data offloading, etc. With those above mentioned WLAN performance features and improvements, new WLAN standard empowered wireless Internet access technologies will enable diverse cost-effective and high-quality Drive-thru Internet applications, ranging from the road safety, trip entertainment to driving efficiency and traffic management.

We close this monograph with three potential future research directions.

(1) IEEE 802.11af based large-coverage Drive-thru Internet. White space spectrum is attractive for its superior broadband quality, longer coverage range, more spectrum resource in rural areas and greater signal penetration. However, white space spectrums are also with variable power transmission constraints and

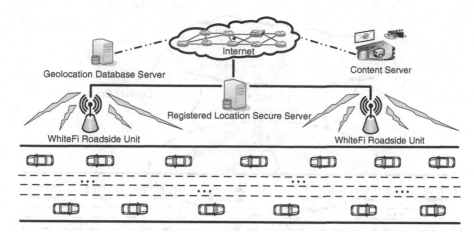

Fig. 5.1 IEEE 802.11af based Drive-thru Internet system model

spatial-temporal variation of available TV channels, which are totally different from ISM band (WiFi band). Hence, how to utilize the white space spectrums for the in-motion and high-quality vehicular Internet access is still unanswered. Shown in Fig. 5.1, a typical vehicular WhiteFi system model is introduced. The vehicular WhiteFi system model is composed of geolocation database server (GDBS), registered location secure server (RLSS) and RSU. Specifically, GDBS provides the local vacant TV white space spectrum query for the dynamic vehicular access. RLSS can be coordinated for the dynamic TV white space spectrum allocation and available white spectrum sharing to optimally utilize the white space spectrum among the neighboring RSUs. RSU is much closer to the vehicular users, which can provide the vehicular users with a long-range Internet access link. For optimally sharing those available white space spectrums, it is crucial to make an in-depth theoretical analysis in dynamic white spectrum sharing scheme, including the calculation of TV white space spectrum enabled RSU coverage range and link capacity. By applying global dynamic optimization theory, how to design the database-based dynamic white space spectrum allocation approach is needed as well, from the perspective of capacity optimization in the whole WhiteFi networking. In addition, how to design a vehicular Internet access protocol for the associated vehicular users in a given RSU coverage is another issue to maximize the vehicular access throughput of whole system while satisfying different QoS levels and application requirements of vehicular users.

(2) **IEEE 802.11ac based Gigabit Drive-thru Internet.** For the investigated IEEE 802.11ac enabled Drive-thru Internet, it supports the single-user MIMO (SU-MIMO) technology for the uplink transmission, and both the SU-MIMO and multi-user MIMO (MU-MIMO) technology for the downlink transmission. In Fig. 5.2, on the one hand, IEEE 802.11ac supports multi-user MIMO (MU-MIMO) technology by which the RSU can transmit multiple independent Gigabit streams simultaneously to multiple mobile devices. However, due to the vehicular

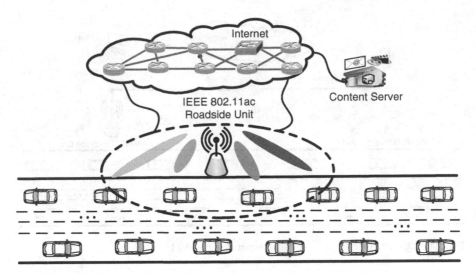

Fig. 5.2   IEEE 802.11ac based Drive-thru Internet system model

high mobility, channel fading characteristic and distinct users' quality of service requirements, how to efficiently and optimally schedule the multiple Gigabit streams to the in-motion vehicles are needed for further study, guaranteeing the fairness of streams scheduling and high-throughput of multi-user multi-gigabit Drive-thru Internet. On the other hand, IEEE 802.11ac technology does not expand the WiFi coverage too much (typically with 160 m coverage radius), for large file transfer, high definition video streaming, even P2P vehicular content sharing on the go, how to enlarge the WiFi connection time and further boost the download performance is crucial for the vehicular Internet access services.

(3) **Cellular/IEEE 802.11ac WiFi/IEEE 802.11af WiFi Interworking for vehicular applications.** IEEE 802.11ac WiFi and IEEE 802.11af WiFi system can provide large-coverage and high-capacity connectivity, respectively, compared with the current IEEE 802.11 WiFi technology which can only be suitable to support the cellular offloading for vehicular applications. Shown in Fig. 5.3, the well-deployed cellular networks, IEEE 802.11ac WiFi and IEEE 802.11af WiFi coexisting scenarios can be helpful to form a cost-effective and high-quality vehicular networking supporting diverse vehicular applications, e.g., road safety, trip entertainment, driving efficiency, and traffic management, etc. However, the dynamic resource availability, strict resource utilization requirements, high vehicular mobility, and the intensive vehicular contending environment will make this type of heterogenous vehicular communication networking challenging. Hence, we need to consider the cellular/IEEE 802.11ac Wi-Fi/IEEE 802.11af WiFi interworking approach for guaranteeing the vehicular service continuity and QoS requirements while balancing the cost issue.

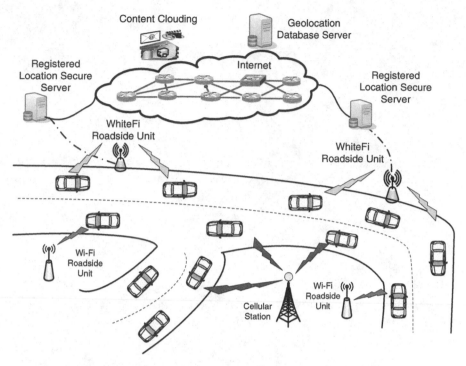

Fig. 5.3 Cellular/IEEE 802.11ac Wi-Fi/IEEE 802.11af WiFi Interworking scenario

References

1. A.B. Flores, R.E. Guerra, E.W. Knightly, P. Ecclesine, S. Pandey, Ieee 802.11 af: a standard for tv white space spectrum sharing. IEEE Commun. Mag. **51**(10), 92–100 (2013)
2. E. Perahia, M.X. Gong, Gigabit wireless lans: an overview of ieee 802.11 ac and 802.11 ad. ACM SIGMOBILE Mob. Comput. Commun. Rev. **15**(3), 23–33 (2011)
3. M. Gast, *802.11 ac: A Survival Guide* (O'Reilly Media, Sebastopol, 2013)
4. R. Van Nee, Breaking the gigabit-per-second barrier with 802.11 ac. IEEE Wirel. Commun. **18**(2), 4–4 (2011)
5. T.H. Luan, L.X. Cai, J. Chen, X. Shen, F. Bai, Engineering a distributed infrastructure for large-scale cost-effective content dissemination over urban vehicular networks. IEEE Trans. Veh. Technol. **63**(3), 1419–1435 (2014)
6. M. Conti, S. Giordano, Mobile ad hoc networking: milestones, challenges, and new research directions. IEEE Commun. Mag. **52**(1), 85–96 (2014)

Printed in the United States
By Bookmasters